The Embodiment of Senses

Babu Thaliath

The Embodiment of Senses

Exploring the Spatiotemporal Structure and Extension of Sensory Perception

Springer

Babu Thaliath
Jawaharlal Nehru University
New Delhi, India

ISBN 978-981-96-8467-0 ISBN 978-981-96-8468-7 (eBook)
https://doi.org/10.1007/978-981-96-8468-7

The original submitted manuscript has been translated into English. The translation was done using artificial intelligence. A subsequent revision was performed by the author(s) to further refine the work and to ensure that the translation is appropriate concerning content and scientific correctness. It may, however, read stylistically different from a conventional translation.

Previously published by Karl Alber Verlag

This Springer imprint is published by the registered company Springer Nature Singapore Pte Ltd.
The registered company address is: 152 Beach Road, #21-01/04 Gateway East, Singapore 189721, Singapore

In memoriam

Olappamanna Damodaran Namboodiripad

Preface

This monograph is a continuation and intended conclusion of an investigation within the framework of the theory of perception, which I introduced in my doctoral thesis more than two decades ago and pursued further during my postdoctoral research in Berlin and Cambridge. It concerns the spatial or bodily and extra-bodily extension of sensory perceptions in general and of vision in particular. The research spanned three years, during which I wrote the monograph as part of my postdoctoral research between 2010 and 2012 at the Department of History and Philosophy of Science at the University of Cambridge.

The spatial extension of sensibility (Sinnlichkeit) remains an unsolved problem in the history of modern philosophy and science to this day. The traditional aporias of sensibility in relation to the bodily and the extra-bodily extension of sensory perceptions—especially the extra-bodily extension of the sense of sight and hearing—can be re-examined in the framework of historiography and philosophy of science. Methodologically, the study presupposes the analogy between the bodily and the extra-bodily extension of sensibility, which in turn refers to a subsequent complementarity between the philosophical-perceptual and scientific foundations of sensibility. The analogicity of sensibility is limited to the spatiotemporal structures of sensory perceptions, represented by their bodily and extra-bodily localisation and temporal simultaneity, while maintaining the ontological differentiation between the sensory qualia—such as colour, sound, pain or the sensation of cold. Likewise, the complementarity between the perceptual-theoretical and scientific foundations of sensibility primarily concerns the spatiotemporal structures underlying all bodily and extra-bodily sensory perceptions. The factum of complementarity should significantly broaden the horizon of research by taking into account and even incorporating the physiological or neuronal basis of the physical expansion of sensibility in the context of a philosophical theory of perception.

On the basis of a few examples already discussed in the early modern history of optics and dioptrics, it is attempted to show how the analogicity of sensibility is based on the complementarity between the perceptual-theoretical and scientific foundations of sensibility. By serving as the basis for the analogicity of sensory perceptions, this complementarity extends the possibility of scientific-methodological analogisation beyond bodily sensibility to the extra-bodily extension of the senses of sight and hearing. Such an extension requires, first and foremost, the analogy between the bodily and the extra-bodily extension of sensibility, whereby the sense of touch, which most closely determines and ensures the spatiotemporal structures of sensibility, acquires a clear priority over all other forms of sensibility. This primacy of touch, which gives the other sensory perceptions their tactile character, is manifested above all in the optical or auditory touch of objects. On the basis of a methodological analogy between the bodily and the extra-bodily sensory perceptions, many prevalent aporias of the spatial and temporal extension of the visual and auditory senses can be re-examined and resolved.

Furthermore, the aporias of sensibility necessarily imply a paradigm shift with regard to the perception-theoretical approach; in the emergence of pre-logical and purely aesthetic *cognition* in sensibility—primarily in vision—the *referential* source or starting point no longer seems to lie in a transcendental subject, but in the real object itself. If such a reversal of epistemological or perception-theoretical referentiality is transferred to scientific methodology, it can solve the traditional aporias of sensibility and thereby produce fruitful scientific knowledge. In this regard, the aporias of visual perception are examined, including the visual perception of size, distance or depth and position, the perspectivity of seeing, optical virtuality, visual perception of free space, and the aporia of the objective localisation of the sense of hearing. The aporias of visual spatial perception, such as the aporia of *object-size-constancy* in relation to visual size perception, as emphasized by Condillac, clearly suggest a reversal of the prevailing transcendental-perceptional referentiality and a subsequent assumption of the direct participation of objects and *their* references in all sensory perceptions. The reversal of the referentiality of the theory of perception and the subsequent accentuation and recognition of the referentiality of the object have the consequence that the spatiotemporal structures of sensibility disengage from the transcendental framework and reclaim the status of the *primary qualia*—as suggested by philosophers of scholasticism and in the early modern period, such as John Locke. The spatiotemporal structures of sensibility, both bodily and extra-bodily, thus form the transcendentally irreducible and as such indestructible *skeleton* of sensually perceivable reality. The purely subjective secondary qualia of sensibility spread on this skeleton like flesh, giving rise to various forms of the real embodiment of sensibility.

The investigation has its origins in my doctoral thesis on Erwin Panofsky's seminal treatise *Perspektive als Symbolische Form*. The idea of pursuing research on the genesis of Renaissance perspective and—further—on the perspectival structure of vision came to me through my participation in a few special and fortunate events: an art history proseminar on *Giotto* with Prof. Dr. Wilhelm Schlink at the Faculty of Art History at the University of Freiburg, a philosophical block seminar on *'Symbolica investigatio'—Philosophieren in Bildern bei Cusanus*, which I attended in the summer semester 1998 with Prof. Dr. Klaus Jacobi in Freiburg, and the postgraduate colloquia with the general topic *Bild und Bildlichkeit* with Prof. Dr. Gottfried Boehm at the Faculty of Art History at the University of Basel, in which I participated for several years (since the winter semester 1998/99). I am very grateful to

Prof. Dr. Wilhelm Schlink for introducing me to the historical genesis of Renaissance perspective and for his continued support of my doctoral thesis. I developed a keen interest in the study of sensibility (Sinnlichkeit)—especially the sense of sight—during my doctoral research between 1998 and 2003 in Freiburg, which I owe above all to my long-standing acquaintance with Prof. Dr. Gottfried Boehm at the Art History Seminar at the University of Basel. I am deeply indebted to Prof. Dr. Klaus Jacobi and Prof. Dr. Gottfried Boehm, the supervisors of my doctorate (Promotion) in Freiburg and Basel, for their encouragement and suggestions and for generously supporting my research. I would also like to thank Prof. Dr. Dominik Perler, Chair of Theoretical Philosophy at Humboldt University in Berlin, for his abiding interest in my research projects and for supervising my post-doctoral research in the field of early modern *mechanical philosophy* over a relatively long period between 2005 and 2013. I also owe my continued engagement with this scientific subject to my long-standing acquaintance with Prof. Dr. Martin Kemp, whom I met in 2000 at the Faculty of Art History at the University of Oxford, as well as the enduring research interest I have gained since then in his basic idea of *structural intuition*.

This monograph has its roots in a preliminary-propaedeutic treatise titled *Die Räumlichkeit der Empfindung* (The Spatiality of Sensation) that I wrote in 2007 during a month-long research stay at the Department of History and Philosophy of Science, University of Cambridge. It was later published as a preprint in 2008 on the Freiburger Dokumentenserver (FreiDok) of the Albert Ludwig University of Freiburg. During that brief but impactful research stay, hosted at Magdalene College in Cambridge, I was generously supported by Prof. Dr. Martin Kusch. I would like to express my sincere gratitude to Prof. Kusch and the Master and Fellows of Magdalene College for their kind support of my research in Cambridge.

Current research builds on the theoretical foundations discussed in the above-mentioned preliminary work and broadens its scope beyond the sense of sight to encompass the full spectrum of bodily and extra-bodily sensory perceptions. I was able to conceive this treatise during my research stay at HPS Cambridge as a Visiting Scholar between 2010 and 2012, and fully conceptualize, write and edit it in parallel with my main work—primary research project on the *Kontextualität der frühneuzeitlichen Wissenschaften* (Contextuality of Early Modern Sciences). I have very fond memories of Prof. Dr. John Forrester, who supervised my research in Cambridge during the first phase. Prof. Dr. Hasok Chang took over supervision during the second phase and has continued to support both my current and future research projects related to the field of early modern history and the philosophy of science. I would like to express my sincere gratitude to Prof. Chang and his colleagues and staff at HPS Cambridge for their ongoing support of my research.

During my research stay in Cambridge, I also made many friends among the local fellows, researchers, teachers and students, who supported me in my research and also in my everyday life in Cambridge. I would like to thank Prof. Tim Crane, Fellow at Peterhouse, for the fruitful and thought provoking philosophical discussions. I would also like to thank all my friends from the *University of Cambridge Newcomers and Visiting Scholars* (NVS) programme and the *Cambridge University India Society* (CUIS) student association, as well as all my friends from the *Postdocs of Cambridge* community. Warm thanks go to Michael O'Sullivan and his wife Moira for their ongoing and very friendly support throughout my research stays in Cambridge.

I am grateful to the Master and Fellows of St. Edmund's College Cambridge for accepting me as a Visiting Scholar. I would also like to express my appreciation to Dr. Anna Gannon FSA, Fellow at St. Edmund's College and John Gannon, Fellow at St. John's College, for their support during my time as a Visiting Scholar at St. Edmund's College Cambridge. Furthermore, I would like to express my gratitude to Dr. Robert Crellin, David Binns, Jenny Hunter, Dr. Jennifer Rampling and Jonathan Rogers for their enduring friendship and support in various academic and extracurricular events and activities in Cambridge. I am also very grateful to the Gerda Henkel Foundation in Düsseldorf, which supported my research in Cambridge and Berlin in various phases and its extensions with a research grant. Without their vital support, this research and the completion of the present monograph would not have been possible.

I am also deeply grateful to my teachers and friends in India and Europe for their continued interest in my postdoctoral research and for all their advice and critical comments: Prof. Wilhelm Schlink, Prof. Anil Bhatti, Helge Naatz, Dr. Amol Kahlon, Dr. Petra Stefanie Vogler, Philipp von Leonhardi, Dr. Sanam Dossal, Vasudevan Alasingachar, Prof. Sundar Sarukkai, John Kottayil, Dr. Silvia De Bianchi, Saurabh Pal, Manjari Mishra, Dr. Inge Anders and Peter Bartke. I would like to thank Dr. Julia Afifi and Julia Engel for carefully proofreading my work. I am profoundly grateful to my wife, Jean Mary, for her unwavering love and support, without which this work would not have been completed. Finally, I would like to thank the Gerda Henkel Foundation in Düsseldorf once again for awarding a printing cost subsidy (Druckkostenzuschuss), which made the publication of this monograph possible.

Cambridge, UK
June 2012

Babu Thaliath

Competing Interests The author has no competing interests to declare that are relevant to the content of this manuscript.

Introduction

Senses alone connect us with our environment. Our access to our own body and the external world happens through various subjective acts—such as feeling, perceiving, cognizing and judging—all of which have their basis in sensibility. As Kant emphasized in the context of his transcendental philosophy, objects are given only in sensibility (Sinnlichkeit); they are cognized by the mind. Accordingly, sensibility bridges the sphere of the subject with that of the object where it acquires a unique ontological status between these different modes of existence. Compared to a logical subject that has the linguistic-conceptual cognition at its disposal, sensibility forms a pre-logical and purely aesthetic domain of the subject, which alone can reach the objects and epistemologically appropriate them.

As never before in the history of philosophy, Kant problematized the epistemological access to objects that are subjectively cognized. The access of the conceptually cognizing subject to the object necessarily requires the connecting function or the mediality of sensibility, which, as such, gains a special transcendental disposition between the conceptual knowledge and the object which is conceptually cognized. In the Kantian system of philosophy, the transcendental refers to an epistemological ascertainment of the direction or referentiality from the subject to the object, that Kant, as is well known, called the Copernican turn in the history of philosophy. In the context of transcendental philosophy, the objects in the cognitive process should direct or orient themselves towards us, i.e., the cognizing subject. Conceptual knowledge arises from the transcendental synthesis which obviously belongs to the sphere of a logical subject. However, the *Transcendental Logic* should be preceded by a *Transcendental Aesthetic*, as the intuition in which alone the objects are given serves as the necessary basis for the conceptual synthesis. Kant therefore begins his *Critique of Pure Reason* with the Transcendental Aesthetic, which, as the doctrine of sensibility, should fix the fundamental guidelines of transcendentalism. That is, the transcendental trait of sensibility should propaedeutically characterize the system of transcendental philosophy.

For this purpose, Kant deploys at the beginning of his main work *Critique of Pure Reason*—in his investigation of sensibility in the context of Transcendental

Aesthetic—the method of epistemological negation and separation of subjective-sensory attributes and qualities from the object.[1] The process of negation and segregation of the sensory qualities characterizes modern epistemologies. Descartes introduced the method of systematic negation in his main works, especially in *Les Meditations Metaphysiques*, in order to demonstrate the existential autonomy of the thinking mind and its complete separation from the body and external, i.e., extra-bodily objects. From the Cartesian method of epistemological negation, the purely spatial expansion—the res extensa—remains as the only objective-residual factum. John Locke extended this notion of objective-residual entity, which cannot be subjectively negated, by ascribing several primary qualities to the object, namely space, time, movement, number and materiality. The objective basis of these primary qualities, which are autonomous from the subject, remains more or less a generally accepted assumption.

Kant's method of negation, as represented in his *Transzendentale Elementarlehre*, surpassed precisely this traditional modern assumption of the objective basis of primary qualities. Within the framework of the *Transzendentale Ästhetik*, space and time are no longer viewed as primary objective qualities, but are ascribed to a transcendental subject as mere a priori representations. As purely a priori notions space and time are, according to Kant, not derived from empirical experiences (a posteriori), but are innate in the transcendental subject. In the context of Transcendental Aesthetic, the "object" can hardly claim its own spatiality and temporality—as its two existential foundations which are completely autonomous from the transcendental subject. For the apriority of space and time relates both to the subjective intuition (*Anschauung*) and to the appearance (*Erscheinung*), which are conceived as such within the strict framework of transcendentalism. The proof for this is the Kantian notion of the thing-in-itself (*Ding an sich*), which constitutes a domain that is completely autonomous from the subject and which, therefore, does not exist in space or time. Through sensibility the subject has access only to the appearance, but not to the thing in itself.

The apriority of space and time is therefore a necessary outcome of the transcendental reduction on the level of sensibility. In the Kantian method of epistemological negation, space and time remain a priori forms of pure intuition, which can also be a priori imagined as mere structures of sensibility. The spatial and temporal extension constitutes the structure or the skeleton of sensibility. If we begin from this residual entity or leftover from the transcendental reduction, we reverse the Kantian method of negation (of all sensory qualities), as a result of which the sensibility builds on a spatiotemporal framework through the addition or filling of the secondary qualities. The spatiotemporal skeleton is inherent in the structure of sensibility.

As evidence for the apriority of space in sensibility, Kant gives examples solely from Euclidean geometry, such as the axiom of the straight line or that of the three-dimensionality of space. Theoretical-geometrical space is obviously an a priori idea of space. Hence the basic idea of the apriority of space within the framework of the

[1] Kant, Immanuel (1990): *Kritik der reinen Vernunft*, ed. Raymund Schmidt, F. Meiner Verlag, Hamburg, p. 64.

Kantian transcendental aesthetic seems to disclose a certain tautology. The empirical or sensible perception of space is actually built on natural spatial structures underlying the individual forms of senses. For example: The perspectival inner structure of the sense of sight, which in contrast to the geometrical-axiomatic intuitions of space is not uniform or constant—or which always changes according to the point of view of the beholder—cannot be ascribed the apriority of a purely geometrical intuition. Rather, the perspectival structure or the perspectivity of seeing, which basically comes about within the framework of a natural geometry by the light rays that are reflected from the objects and converge on the eye point, seems to be *given* empirically. Likewise, the apriority of the natural bodily and extra-bodily extension and localization of other sensations seems equally questionable; Their spatiality turns out to be quite heterogeneous—in accordance with the different forms of sensibility and especially in relation to the perspective-structured seeing.

The original heterogeneity of the various sensory perceptions—seeing, hearing, touch, taste, smell and bodily sensations such as pain or cold—and their spatial structures were strategically subsumed under the general concept of intuition (*Anschauung*) within the context of the Kantian transcendental aesthetic. Such a subsumption of senses under an apparently meta-empirical concept such as intuition made it possible for Kant to declare space and time as general and pure a priori forms of sensibility. If, on the other hand, the fundamental heterogeneity of the individual sensory perceptions and their spatiotemporal structures are taken into account, this could shake the apriority of space and time as pure forms of sensibility, or even invalidate them.

When Kant emphasizes within the framework of the transcendental aesthetic that space and time cannot be subjectively negated in perception in relation to all sensory qualities and therefore cannot be separated from the sensibility, he tacitly ascribes to space and time as pure forms of intuitions an ontological status analogous to the primary qualities. For Kant derives the transcendental-aesthetic irreducibility and the resulting residuality of space and time undoubtedly from the prevailing modern notion of primary qualities and their objective-existential autonomy from all secondary qualities. Nevertheless, Kant does not ascribe any *objective* status to these primary qualities and their origin in sensibility; they are defined in the strict framework of transcendentalism as pure a priori ideas (*Vorstellungen a priori*). Such apriorization of space and time, however, does not enable Kant to accomplish a completely transcendental reduction of space and time. The ontological status of space and time as primary qualities seems to be retained in the transcendental-aesthetic reduction.

Are space and time mere a priori ideas, as Kant determines them in the strict framework of his transcendental philosophy? The persistence of space and time as primary qualities arouses the suspicion that space and time derive their ontological irreducibility originally from an objective domain that is autonomous from the transcendental subject. Since sensibility alone can form the nexus between the subject and the world—in cognitive as well as perceptual processes—such questions can be posed within the framework of transcendental aesthetic: whether space and time form the indestructible skeleton of reality experienced through our senses, and whether they are absorbed as such—with necessary modulations—in the subjective sensibility? This preferential position of space and time in the *Transcendental Aesthetic* is the result of the transcendental-aesthetic irreducibility of these forms of sensibility, which Kant himself demonstrates in his method of epistemological negation. The irreducibility of space and time also points to their structural unity and finality, which serve as a common basis for the various forms of sensibility. The temporal simultaneity of different sensory perceptions, such as taste, touch, pain and cold, in their bodily extension shows the uniform and irreducible spatiotemporal structures in the bodily extension of sensibility. This also applies to the extra-bodily sensations such as seeing and hearing.

This unity and generality of the spatial and temporal structures guarantee the analogy between different forms of bodily and extra-bodily sensibility. In the bodily extension and temporal simultaneity, physical sensations prove to be analogous to one another. Different bodily sensory perceptions, such as pain, cold, softness and taste, clearly show in their bodily localization, extension and in their temporal simultaneity analogous spatiotemporal structures inherent in all forms of bodily sensibility. However, such an analogy between bodily sensations necessarily arises from a material phenomenality that they share with one another, namely the neural causality of the bodily sensibility. Both the sensory qualities and their bodily localization, extension and simultaneity owe their origin and duration to the neural network that is spread throughout the body and is controlled by the brain.

This purely scientific basis of the spatial and temporal structure of bodily sensibility seems to refute from the outset the traditional transcendental-philosophical claim to the apriority of space and time in the domain of sensibility. However, on closer inspection, the purely subjective and the purely bodily emergence of sensory perceptions do not turn out to be contradictory, but necessarily complementary. What prevents us, however, from thinking or speculating a complementarity between the subjective and the material-bodily origin of sensibility is our tendency towards priority-thinking to which we have long become accustomed in both philosophy and natural sciences.

Now we should investigate whether the above-discussed analogy between the bodily sensory perceptions extends beyond the body into the extra-bodily sensibility. Here we are confronted with the problem of drawing analogies between bodily and the extra-bodily sensations. The spatial and temporal extension of extra-bodily senses—apart from the uniqueness and otherness of the perspective spatial structure of vision—seems to legitimize and safeguard the analogy between the bodily and the extra-bodily sensibility. The scientific basis of such an analogy—and of the external extension of senses in general—seems to pose a clear challenge. In order to master this challenge, we have to rely on the above-mentioned complementarity between the subjective and the purely objective considerations of sensibility.

The purely subjective aspect of the bodily and extra-bodily sensibility relates to the emergence of the sensory qualities, such as colour or brightness of the objects in vision, different voices or noises in auditory sensation, the localized tastes on the tongue and the tactile bodily sensations such as warmth or cold. These sensory qualities are secondary qualities. As previously discussed, the spatial extension and localization of these sensory qualities as well as their temporal simultaneity refer to the primary qualities of sensibility. The question now arises as to whether the subject participates in the emergence of the primary qualities of sensibility. From a purely scientific point of view, it becomes clear to us that only the material body, i.e., the neural effectuation in the nervous system spread throughout the body ensures the physical localization, extension and simultaneity of all bodily sensations. That is to say, it is the objective body, which—as opposed to the solely subjective origin of all secondary sensory qualities—takes part directly in the spatial localization, extension and temporal simultaneity of sensory perceptions within the body. While such a participation of the objective body in the bodily-sensory perceptions seems plausible, the search for an analogous participation of the objective-material phenomena in the spatiality and temporality of the external sensory perceptions gives us great difficulties.

Emergence of the secondary sensory qualities commonly suggests an aporia—regardless of whether the phenomenon of sensibility is examined in the context of natural sciences or in the context of philosophical epistemology. This aporia of sensibility arises from the puzzling causal nexus between a purely mental effect and its purely material causation. The various sensory qualities are entirely mental phenomena whose physiological or neural causations are restricted to the sphere of matter and do not consequently transcend its ontological limit. The emergence of the subjective sensory qualities is therefore clearly an ontological leap from a material sphere of causation into a *real* but immaterial mental sphere, which eludes our scientific-rational apprehension.

The complete leap from a material causality to a mental reality, as represented in the emergence of sensory qualities, indicates a causal principle that is primarily of ontological nature. The emergence of the sensory qualities is therefore the consequence of an ontological causality where the elementary modes of being always produce a more complex yet uniform mode of being. The aporia of the emergence of sensory qualities can ultimately be traced back to the ontological causality that underlies every phenomenal and mental emergence. In case of the material causality of the mental phenomena, the aporia of emergence reveals itself in full splendor and clarity.

The emergence of the secondary sensory qualities refers to the direct involvement of the objective phenomena in the emergence of these qualia—in particular, all neuronal states and processes. However, the aporetic nature of this emergence is not easily explained by the demonstrable participation of the objective phenomena; Instead, such participation—on a material-objective level of causation—gives the mental emergence of the sensory qualities the clear status of an unresolvable aporia. The assumption of a direct participation of the objective phenomena in sensory processes of perception, on the other hand, resolves all aporias regarding the origin of the primary qualities of sensibility, as represented both in the bodily as well as in the extra-bodily sensory perceptions. All the prevailing aporias of vision—such as the visual perception of size, distance and position, visual virtuality in dioptric phenomena such as reflection and refraction or the emergence of the perspectival structure of vision—seem to suggest the direct involvement of the objective phenomena in the subjective sensibility.

This necessity that the *factum of object* should be included in subjective sensibility—both from the philosophical and from the purely scientific point of view—was already indicated in the primacy or preferential position of the sense of touch within the framework of medieval scholastic philosophy. Of all the forms of sensory perception, it is the touch that connects the perceiving subject most closely with the perceived object. While seeing necessarily requires a distance from the objects seen, physical touch, which includes many physical sensory perceptions such as pain, warmth, cold or taste, requires that this distance be overcome. The question of the dichotomy of objective existence—namely how things exist as both real and mental at the same time—that was debated in depth in the late medieval scholasticism and the early modern period that followed, relates primarily to the sense of sight.[2] When seeing things directly, images of the objects emerge that do not have materiality compared to the real objects. Such a sensation is hardly possible in the case of direct touch, as the touch or tactile sensation primarily requires the materiality of the object touched. Likewise, the bodily sense of taste, which in its mode of effectuation comes close to the sense of touch (since the tongue should initially feel the objects), determines the materiality of the tasted objects. Accordingly, a mental object arises more from seeing; chewing or swallowing a mental object, on the other hand, does not seem conclusive.

In her work *Zwei Untersuchungen zur nachscholastischen Philosophie* Anneliese Maier discusses the primacy of the sense of touch over all other forms of sensory perception in the medieval-scholastic philosophy and the reversal of this hierarchy of senses in the Cartesian early modernity.[3] The scholastic philosophy established the primacy of the sense of touch by interpreting every other form of sensation as a sensory touch. Colours are purely subjective perceptions, but for the scholastic philosophers they are perceived in the object. The direct participation of the object in sensory perception implies the existential dependence of the secondary sensory qualities on the primary qualities (here: on the spatial distance and the extension of objects), as Anneliese Maier observes:

> Wie die Qualitäten im einzelnen von den primären abhängen sollen, wird, besonders wenn es sich auch um die nicht-taktilen handelt, in der älteren Philosophie nur sehr undeutlich gewusst und gesagt. Die Argumentation geht häufig über die Vorzugsstellung des Tastsinns, denn der ist zwar nicht der vornehmste, aber der notwendigste Sinn, der von allen vorausgesetzt wird, selbst aber keinen voraussetzt.[4]

---

[2] See Perler, Dominik (2004): *Inside and Outside the Mind – Cartesian Representations Reconsidered, in: Perception and Reality. From Descartes to the Present*, ed. Ralph Schumacher, Mentis Verlag, Paderborn, p. 76–77.

[3] See Maier, Anneliese (1968): *Zwei Untersuchungen zur nachscholastischen Philosophie*, Rome, p. 18.

[4] Ibid.

The extension of senses in general and the extra-bodily extension of the sense of sight and hearing in particular can therefore hardly be reduced to the apriority of the forms of intuition, space and time, within the transcendental system of philosophy. The differentiation between the objective-primary and the subjective-secondary qualities of sensory perceptions, as considered in the framework of medieval-scholastic philosophy and later re-established by Locke in early modernity, seems to hold for the origin and existence of all forms of sensibility. The transcendental irreducibility and the subsequent primacy of the primary sensory qualities, which can now be conceived as independent of the perceiving subject, should therefore establish the spatiality and temporality of sensory perceptions as a skeleton of the given reality. The secondary sensory qualities spread over this skeleton like flesh, forming thus the *real* embodiment of senses.

The external extension of the sense of sight and hearing spawns one of the most important aporias of sensibility, the historical persistence of which can be traced back to their insolubility—or to the futile attempts at finding a solution from a prevailing transcendental-subjective point of view. Several individual aporias of the visual perception of space, time and movement can be subsumed under these superordinate aporias. In addition to the previously mentioned aporias of the external extension of the sense of sight and hearing, some of the well-known aporias that have been handed down since antiquity, such as the *moon illusion*, can be re-examined in this context. The aporicity of these perceptual-theoretical problems, which have not yet been resolved, seems to require an equally perceptual-theoretical reversal of the point of view of investigation. Consequently, the conventional and habitual persistence on a purely subjective point of view could be abandoned and replaced by a point of view based on the object. This would legitimize the above-mentioned inevitable complementarity between the philosophical and the scientific investigations. Every aporia ascertained philosophically or in terms of the theory of perception should therefore necessarily and speculatively suggest an objective factum, which requires the investigation of this factum in the context of scientific research. In short: the unsolvable aporicity of the perceptual-theoretical problems briefly outlined above should lead *methodologically* to the apodicticity of the phenomena, that are conceived speculatively at present, but which can be proven objectively and scientifically.

When aporias of sensibility arise through the hurdles of the theory of perception, these riddles in thinking basically point to a subjective impossibility. Such a subjective state then presupposes a necessary search for the problem and its solution in the object (that Aristotle suggests propaedeutically in his book of aporias in *Metaphysics*[5]). This could result in the speculative—and indeed scientific-speculative—solution that scientific investigation has to deal with. Such a methodological shift implies a paradigm shift in research, which, however, does not refute the fundamental complementarity between philosophical and scientific investigation, but supports it instead.

---

[5] See Jacobi, Klaus (2008): *Kann die Erste Philosophie wissenschaftlich betrieben werden? Untersuchungen zum Aporienbuch der aristotelischen »Metaphysik«, in Metaphysisches Fragen.* Colloquium über die Grundform des Philosophierens, eds. Paulus Engelhardt und Claudius Strube, Böhlau Verlag, Collegium Hermeneuticum, Bd. 12, Köln – Weimar – Wien, p. 31 ff.

Accordingly, the emergence of an aporia in the course of *subjective*-philosophical research should necessitate the scientific speculation of an objective factum as an indispensable solution. The attempt to experimentally prove this objective factum would advance and significantly enrich the fundamental scientific research.

Babu Thaliath

References

Aristotle (1995). *Metaphysics*. In: *The complete works of Aristotle*. Jonathan Barnes ((Ed.). Vol. II. Princeton University Press. Princeton.

Jacobi, Klaus (2008). *Kann die Erste Philosophie wissenschaftlich betrieben werden? Untersuchungen zum Aporienbuch der aristotelischen »Metaphysik«*. In: *Metaphysisches Fragen. Colloquium über die Grundform des Philosophierens*. Paulus Engelhardt and Claudius Strube (Eds.). Böhlau Verlag. Collegium Hermeneuticum. Bd. 12. Köln – Weimar – Wien.

Kant, Immanuel (1990). *Kritik der reinen Vernunft*. Raymund Schmidt (Ed.). F. Meiner Verlag. Hamburg.

Maier, Anneliese (1968). *Zwei Untersuchungen zur nachscholastischen Philosophie*. Rom.

Perler, Dominik (2004). *Inside and Outside the Mind – Cartesian Representations Reconsidered*. In: *Perception and Reality. From Descartes to the Present*. Ralph Schumacher (Ed.). Mentis Verlag. Paderborn.

Contents

Chapter 1
The Transcendental Status of Sensibility

1.1 The Spatiotemporal Structure of Sensibility

As is generally known, cognition is based on a bond between the knowing or cognizing subject and the cognized object. As an epistemological nexus, every cognition includes these two essentially different modes of existence or being. Accordingly, the mode of cognition primarily forms a subjective construct—in the form of linguistic terms and expressions or extra-linguistic signs—that disguises the purely phenomenal reality like a veil. As such, the knowing subject is primarily a logical subject that produces knowledge in a logical-synthetic nexus with objects. The mode of the epistemological bond between the subject and the object can hardly be defined as a composition, as the logical abstraction and the processes of synthesis give rise to a modal and formal separation between knowledge or cognition and the object that is cognized. Strictly speaking, all knowledge—as constructs of a logical subject—belong to an extra-phenomenal domain of the subject. Every cognition presupposes the direct or indirect *presence* of the cognized object. However, the *objective* content of the cognition seems to be conceived in a more metaphorical sense.

In contrast to the logical-synthetic cognition, sensibility, in which alone the objects are *given*, is regarded as the necessary preliminary stage of the cognitive process. In the Kantian transcendental system of epistemology, the domain of sensibility is more or less hierarchically subordinated to a logical subject that synthesizes the objects given in sensibility into conceptual knowledge. The Kantian idea of intuition (Anschauung) relates in principle to all modes of sensibility—vision, hearing, touch, smell and taste—the spatial and temporal *extension* of which is determined not just by phenomenon, but also and primarily by a transcendental subject. The givenness of spatially and temporarily extended objects in sensory perception seems to arise in the mode of annexation, the compositional unity of which is based on the irreducible facticity of space and time, which constitute the basic mode of the existence of the objects perceived and also of the sensory perception itself. Apart from this compositional unity, which incorporates the spatial and temporal extension of the subjective

B. Thaliath, *The Embodiment of Senses*, https://doi.org/10.1007/978-981-96-8468-7_1

sensibility and the objects, sensibility *qualitatively* consists of different modes of sensory nexus between the subject and the object, such as colour, sound, warmth, softness, smell, taste, etc. These and similar *objects* of pure sensory perception belong to a pre-logical domain of the subject, which we justifiably call the domain of an aesthetic subject. In comparison to the linguistic-conceptual knowledge, the realm of sensibility builds on a pre-logical and purely aesthetic-synthetic nexus between the subject and the phenomenon or object.

Conceptual synthesis or conceptual-synthetic judgement is—in the Kantian transcendental system—clearly a function of the logical subject and forms as such the main subject of a *Transzendentale Logik*. The transcendental logic should necessarily be preceded by a transcendental aesthetic that mainly examines the formal structure of sensibility. Intuition (Anschauung) is the key concept that Kant introduces and discusses in detail in the propaedeutic doctrine of the transcendental aesthetic. Kant ascribes to perception a function of mediation or nexus; Only through perception the understanding (Verstand), which cognizes *conceptually*, is connected to the objects given in the empirical intuition. The epistemological mediation of perception alone enables the subject to relate or *refer* to the object (Gegenstand) in conceptual cognition:

> Auf welche Art und durch welche Mittel sich auch immer eine Erkenntnis auf Gegenstände beziehen mag, so ist doch diejenige, wodurch sie sich auf dieselbe unmittelbar bezieht, und worauf alles Denken als Mittel abzweckt, die Anschauung. Diese findet aber nur statt, so fern uns der Gegenstand gegeben wird; dieses aber ist wiederum, *uns Menschen wenigstens*, nur dadurch möglich, daß er das Gemüt auf gewisse Weise affiziere. Die Fähigkeit (Rezeptivität), Vorstellungen durch die Art, wie wir von Gegenständen affiziert werden, zu bekommen, heißt Sinnlichkeit. Vermittelst der Sinnlichkeit also werden uns die Gegenstände gegeben, und sie allein liefert uns Anschauungen; durch den Verstand aber werden sie gedacht, und von ihm entspringen Begriffe. Alles Denken aber muß sich, es sei geradezu (directe), oder im Umschweife (indirecte), *vermittelst gewisser Merkmale*, zuletzt auf Anschauungen, mithin, bei uns, auf Sinnlichkeit beziehen, weil uns auf andere Weise kein Gegenstand gegeben werden kann.[1]

The mediation of intuition proves to be a gradual process that is subject to an epistemological order. Accordingly, various objective and subjective entities on which perception is based are defined and integrated into the process of perception. In contrast to the understanding and its a priori-conceptual synthesis, the perception process does not begin with its a priori-formal structure of space and time, but with

[1] Kant (1990, p. 63).
See Kant, Immanuel (1998): *Critique of Pure Reason*, transl. & ed. Paul Guyer and Allen W. Wood, Cambridge University Press, Cambridge, p. 155 (A19–20; B33–34).
"In whatever way and through whatever means a cognition may relate to objects, that through which it relates immediately to them, and at which all thought as a means is directed as an end, is intuition. This, however, takes place only insofar as the object is given to us; but this in turn, is possible only if it affects the mind in a certain way. The capacity (receptivity) to acquire representations through the way in which we are affected by objects is called sensibility. Objects are therefore given to us by means of sensibility, and it alone affords us intuitions; but they are **thought** through the understanding, and from it arise concepts. But all thought, whether straightaway (*directe*) or through a detour (*indirecte*), must ultimately be related to intuitions, thus, in our case, to sensibility, since there is no other way in which objects can be given to us."

the object and its givenness through sensibility (Sinnlichkeit). First of all, the object *affects* the subject or the mind (Gemüt). Subject's receptive ability to be affected by objects then forms sensibility, whereby the objective effect itself constitutes a sensory perception. *Empirical* intuition relates to the object through sensory perception. Kant defines the object *given* in empirical intuition as *appearance* (Erscheinung).[2] In this way, the process of intuition occurs in a step-by-step progression, namely in a progression from (1) the object, (2) its affecting the mind in the form of a sensory perception and the (3) reception of this objective effect in the subjective faculty of sensibility up to (4) the integration of the objects given by means of sensibility (as appearances in empirical intuitions) in an a priori formal structure of space and time.

In this context it is striking to note that Kant differentiates the (a priori) participation of the subject from the (a posteriori) participation of the object already in a preliminary stage of this process of intuition. In empirical intuition, the object is *not merely given*; the (conceptually) indeterminate object of empirical intuition is an *appearance* that is composed of matter (a posteriori) and form (a priori). The matter of the appearance *corresponds to the sensation* and should therefore be *given first*, whereas the form of the appearance, in which the manifold of the appearance can be arranged in certain relations, "must a priori exist in the mind."[3] Likewise the intuition, through which the understanding cognizes the objects conceptually, is not conceived in a uniform, but rather in a compositional mode. Kant postulates a pure intuition, that is the pure form of sensibility, and distinguishes it from the empirical intuition, which *contains* the sensory perceptions or the matter of the given appearances. Only through a complete *elimination* of all intelligible and sensory attributes from the empirical intuition can pure intuition be isolated and its two pure a priori forms, namely space and time, identified.

In order to fully extract the pure a priori forms of sensibility, Kant uses the method of systematic negation and separation—a method which, in the context of modern epistemology, can be traced back to the Cartesian method of doubt. Pure intuition consists of the two a priori forms of sensibility, namely of space and time. The methodical isolation of the pure a priori forms of sensibility corresponds, at the same time, to the dismantling of the synthetic nexus between a sensibly perceiving and conceptually cognizing subject and the perceived and cognized object. First of all, the logical synthesis in objective knowledge is deconstructed, so that the sensibility alone remains. After that, all components of the sensibility are isolated so that space and time remain as two pure forms of sensory perception and, at the same time, of appearance. This second segregation attains an epistemological

---

[2] Kant (1990, p. 63). "Die Wirkung eines Gegenstandes auf die Vorstellungsfähigkeit, so fern wir von demselben affiziert werden, ist Empfindung. Diejenige Anschauung, welche sich auf den Gegenstand durch Empfindung bezieht, heißt empirisch. Der unbestimmte Gegenstand einer empirischen Anschauung heißt Erscheinung."

See Kant (1998, p. 155, B24; A 20).

"The effect of an object on the capacity for representation, insofar as we are affected by it, is **sensation**. That intuition which is related to the object through sensation is called **empirical**. The undetermined object of an empirical intuition is called **appearance**."

[3] Kant (1990, p. 64).

finality where space and time appear as irreducible residual entities. However, the epistemological irreducibility and finality of these residual entities is ultimately based on their ontological irreducibility and finality—a fact that can be derived from both the Cartesian and the Kantian method of systematic negation and segregation.[4]

If we reverse this systematic segregation of the pure forms of sensibility (by first considering the finality of the pure forms, namely space and time, and their function in the process of sensation, perception and cognition) or, in other words; if we start from these residual and as such irreducible entities, which form the irreducible basis of perception, and proceed to the synthetic construction of our sensibility, we recognize these forms as necessary structures of our senses. All modes of our sensibility necessarily arise in the spatial and temporal form and structure, to which Kant only ascribes an a priori status. In the construction of sensibility, the sensations, initiated by objects, provide the raw material, while space and time as pure a priori forms structurally design the construct of the senses. The apriority of these forms is a necessity according to Kant which is referred to several times in the context of a detailed treatment of space and time in the *Transcendental Aesthetic*. This is clearly demonstrated in all axiomatic geometrical cognitions. In order to demonstrate the apriority of space, Kant repeatedly cites a few examples from Euclidean geometry, such as the axiom of the straight line or that of the three-dimensionality of Euclidean space.

That space has three dimensions is necessarily a priori knowledge. The apodictic certainty of this geometric principle and the possibility of its a priori construction are based on this a priori knowledge.[5] This axiomatic knowledge points to the structure of the Euclidean space, which forms the *background* of all geometrical forms and laws. In the *Transcendental Aesthetic*, examples of the a priori of space are exclusively from the Euclidean geometry. The Euclidean-geometric space and its structure, as represented primarily in its three-dimensionality, corresponds to the reality of our living space and the Newtonian absolute space of the cosmos, yet it is an a priori conception of space, i.e., a theoretical space that we let surface in our productive

---

[4] Ibid., p. 65. „In der transzendentalen Ästhetik also werden wir zuerst die Sinnlichkeit isolieren, dadurch, daß wir alles absondern, was der Verstand durch seine Begriffe dabei denkt, damit nichts als empirische Anschauung übrigbleibe. Zweitens werden wir von dieser noch alles, was zur Empfindung gehört, abtrennen, damit nichts als reine Anschauung und die bloße Form der Erscheinungen übrigbleibe, welches das einzige ist, das die Sinnlichkeit a priori liefern kann. Bei dieser Untersuchung wird sich finden, daß es zwei reine Formen sinnlicher Anschauung, als Prinzipien der Erkenntnis a priori gebe, nämlich Raum und Zeit, mit deren Erwägung wir uns jetzt beschäftigen werden.“

See Kant (1998, p. 157, A22).

“In the transcendental aesthetic we will therefore first isolate sensibility by separating off everything that the understanding thinks through its concepts, so that nothing but empirical intuition remains. Second, we will then detach from the latter everything that belongs to sensation, so that nothing remains except pure intuition and the mere form of appearances, which is the only thing that sensibility can make available *a priori*. In this investigation it will be found that there are two pure forms of sensible intuition as principles of *a priori* cognition, namely space and time, with the assessment of which we will now be concerned.”

[5] Kant (1990, p. 67).

imagination. In contrast, the *space of sensibility* turns out to be hardly uniform. The perception of space is conveyed to us mainly through visual, haptic and auditory perceptions. Among these, the visual perception of space largely constitutes both our direct experience of an aesthetic space and our (a priori) notions of the geometric forms and structures. However, the visual space, that is, the space *seen*—regardless of whether it is experienced directly by looking or is simply constructed in the productive imagination—is *necessarily* a perspectival space. In other words: the immediate visual space (a posteriori) and the a priori imaginary space are based on the structure of perspective in an *undifferentiated* manner. The above-mentioned geometrical axiom requires, in addition to its real aesthetic experience or a priori imaginability, a certain logical synthesis. The pure visibility of the three-dimensionality of Euclidean space proves to be *modally uniform* both in the direct visual experience and in the a priori imagination, that is, it is subject to the irreducible structure of perspectivity. Whether we really represent the three-dimensionality of space in the form of spatial coordinates or just imagine them (a priori), we are only able to *perceive* them in a perspectival spatial structure (in that some angles at the origo—the zero point—that are right angles in reality, are seen in a perspectival distortion, as acute and obtuse angles).

The structure of the vision—both in the a posteriori visual experience and in the a priori imagination—corresponds more to a perspectival space; the geometric space as well as the geometric shapes and structures seem to be abstracted from this original aesthetic space. In the context of *Transcendental Aesthetic*, and in particular in the context of the investigation of the apriority of the concept of space, Kant does not attempt to imagine an aesthetic space—as a direct or concrete space of sensibility—and separate it from the theoretical-abstract space of geometry. This becomes particularly clear when we observe that all geometrical-axiomatic notions (a priori) necessarily presuppose the visualizations that are and remain subject to a perspectival spatial structure. In order to distinguish between the real perspectival spatial structure of the immediate visual experience and the a priori imagination of the geometric space, one must isolate vision from the other modes of sensibility. Kant, on the other hand, subsumes all individual modes of sensibility within the strict framework of a *transcendental* aesthetic under a basic conception of intuition (Anschauung). Anyhow, the concept of perception refers primarily to a visual space, that is, to seeing.

1.2 The Apriorization of Space and Time

„Der Raum ist eine notwendige Vorstellung a priori, die allen äußeren Anschauungen zugrunde liegt".[6] How does Kant define external intuitions? They relate unconditionally to the sensibility through which alone we can perceive the extension of our body and the extra-bodily space in which the external objects are located:

> Vermittelst des äußeren Sinnes, (einer Eigenschaft unseres Gemüts), stellen wir uns Gegenstände als außer uns, und diese insgesamt im Raume vor. Darinnen ist ihre Gestalt, Größe und Verhältnis gegen einander bestimmt, oder bestimmbar. Der innere Sinn, vermittelst dessen das Gemüt sich selbst, oder seinen inneren Zustand anschauet, gibt zwar keine Anschauung von der Seele selbst, als einem Objekt; allein es ist doch eine bestimmte Form, unter der die Anschauung ihres inneren Zustandes allein möglich ist, so, daß alles, was zu den inneren Bestimmungen gehört, in Verhältnissen der Zeit vorgestellt wird.[7]

It is the sensibility alone that connects the subject with the *real* body and with the extra-bodily objects. As external intuitions, all modes of our sensibility are spatially extended. The spatiality, on which our external or bodily and extra-bodily sensibility is based, constitutes, according to Kant, a *necessary a priori notion.* As already stated earlier, Kant cites in the *Transcendental Aesthetic* the Euclidean-geometrical axioms as examples for the apriority of the conception of space. According to Kant the apodictic certainty of the geometric axioms is based on the necessary apriority of space. This seems to be a special strategy within the framework of his doctrine on the *transcendental* aesthetic. The space of geometry is *primarily* a space of representation, which—as Kant discusses in detail in "Axiome der Anschauung" (Axioms of Intuition)—we construct in our productive imagination. As such it is clearly an a priori notion of space. In contrast, the space that underlies our sensory perceptions—especially vision—as well as the *external* objects perceived through external senses, is necessarily an aesthetic space that differs from the a priori imaginary space. The question now arises as to whether the spatiality of our immediate sensory perceptions and of the objects perceived by the senses is a mere a priori notion or primarily an a posteriori givenness (Gegebenheit)?

Kant conceives space and time within the strict framework of his *transcendental* system of philosophy. The Transcendental Aesthetic, which as the first part of the *Transzendentale Elementarlehre* forms a propaedeutic to his transcendental philosophy, is directed from the outset to the justification of the apriority of space and

[6] Kant (1990, p. 67). See Kant (1998, p. 158, A24). „Space is a necessary representation, a priori, which is the ground of all outer intuitions."

[7] Kant (1990, p. 66).
See Kant (1998, p. 157, A23; B37–38).
"By means of outer sense (a property of our mind) we represent to ourselves objects as outside us, and all as in space. In space their form, magnitude, and relation to one another is determined, or determinable. Inner sense, by means of which the mind intuits itself, or its inner states, gives, to be sure, no intuition of the soul itself, as an object; yet it is still a determinate form, under which the intuition of its inner state is alone possible, so that everything that belongs to the inner determinations is represented in relations of time."

time. This strictly transcendental approach seems to take on the trait of a contextualization, i.e., a philosophical-historical contextualization. Space and time are—as purely a priori forms of sensibility—ascribed to a perceiving subject rather than to the phenomenal reality that is perceived through senses. Indeed, Kant does not negate the spatial and temporal existence of the objective world, still he is strategically silent on this important matter in the *Transcendental Aesthetic.* If the world of objects (to which our bodily existence belongs) truly exists independent of the sensuously perceiving and conceptually cognizing subject, we can, adding to Kant, state that space and time are necessary a priori notions *only within the context of the transcendental philosophy*. However, at the beginning of the *Transcendental Aesthetic*, Kant poses the question of the true ontological determination of space and time between their mere apriority or a priori representation and their a posteriori givenness:

> Was sind nun Raum und Zeit? Sind es wirkliche Wesen? Sind es zwar nur Bestimmungen, oder auch Verhältnisse der Dinge, aber doch solche, welche ihnen auch an sich zukommen würden, wenn sie auch nicht angeschaut würden, oder sind sie solche, die nur an der Form der Anschauung allein haften, und mithin an der subjektiven Beschaffenheit unseres Gemüts, ohne welche diese Prädikate gar keinem Dinge beigelegt werden können?[8]

Kantian answers, on which his *Transcendental Aesthetic* and furthermore his Transcendental Philosophy in general are based, are obviously a yes to the first and a no to the last question. Further with this question and its discussion in the Transcendental Aesthetic, the contextuality of the Transcendental Philosophy is tacitly established.

The problem of sensory—especially visual—perception of space has a pre-Kantian history of development in the modern philosophy. In *Der Mensch* (Traité de l'homme), Descartes examines the development of the visual perception of space, i.e., the visual perception of size, distance and position of objects. The Cartesian investigation of visual spatial perception is based on the relationship between the visual and the tactile sensations.[9] Whether the mere tactile sensation can suggest visual spatial perception was problematized in the well-known Molyneux question raised in seventeenth century. The question posed by William Molyneux prompted John Locke, and also George Berkeley, to examine in detail this apparently unsolved problem in the visual perception of space. Berkeley's *An Essay towards a New Theory on Vision* dealt exclusively with the phenomenon of vision, where to a large extent the perception of the visual space, as represented in the visual perception of size,

---

[8] Kant (1990, p. 66).
See Kant (1998, p. 157, B38).
"Now what are space and time? Are they actual entities? Are they only determinations or relations of things, yet ones that would pertain to them even if they were not intuited, or are they relations that only attach to the form of intuition alone, and thus to the subjective constitution of our mind, without which these predicates could not be ascribed to any thing at all?"

[9] In the part *Der Mensch* (Traité de l'homme) in his main work *Die Welt* (Le Monde), Descartes examines the involvement of the sense of touch in the sense of sight, i.e., in the visual perception of position, shape, distance, size, etc.: See Descartes, René (2015): *Die Welt*, trans. and ed. by Christian Wohlers, Felix Meiner Verlag, Hamburg, p. 241. Descartes then explains the involvement of the sense of touch in the sense of sight on the basis of various demonstrations, such as the following (Fig. 1.1).

position and distance, as well as the perceptual-theoretical divide between visual and tactile sensations was examined. The Molyneux question and its discussion by Locke and Berkeley met with considerable response in France—from Diderot, Condillac and Voltaire. It is worth mentioning in this regard Condillac's study *Traité des Sensations* (1754) in which it is clearly stated to what extent the visual perception of size and distance remains an unsolved problem—despite all the attempts to solve it in the history of modern science and philosophy.[10]

In addition to these philosophical-epistemological investigations, geometrical optics developed into an autonomous field of science in the early modern period. This

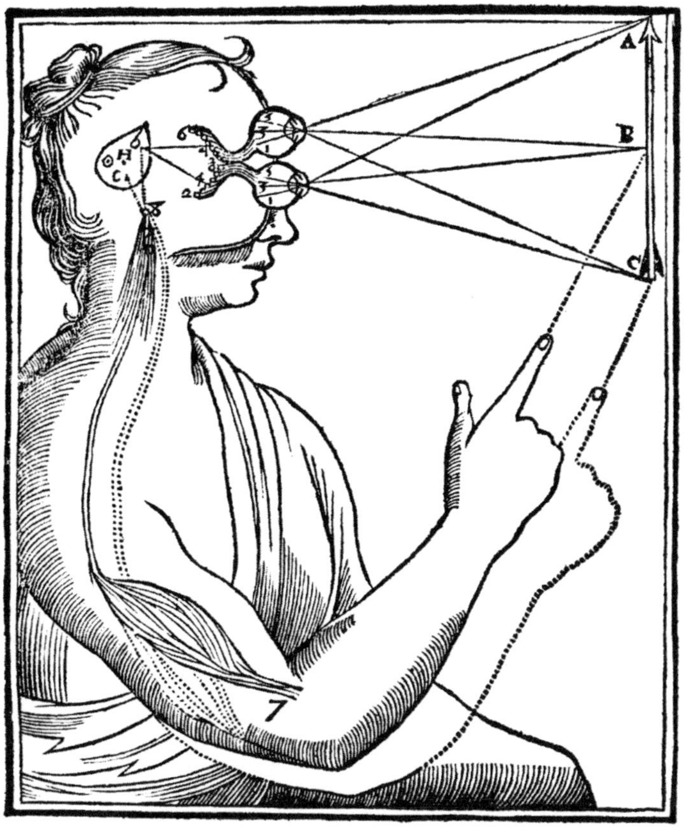

Fig. 1.1 Descartes' *Traité de l'homme*: The involvement of the sense of touch in the sense of sight. Descartes, René (2015): *Die Welt*, trans. and ed. Christian Wohlers, Felix Meiner Verlag, Hamburg, p. 287

Ibid., p. 287.

[10] The renowned psychologist Michael J. Morgan published a treatise in 1977 on the Molyneux question and its reception in the early modern age by Locke, Berkeley, Diderot, Condillac, Voltaire and others, as well as the topicality of this problem in current research on visual space perception. Morgan, Michael J. (1977): *Molyneux's Question. Vision, Touch and the Philosophy of Perception*, Cambridge.

period also saw the discovery of the central perspective (by Filippo Brunelleschi) and its establishment in the art of painting of the Renaissance, as well as its influence on the early modern theories of vision. Both the modern geometrical optics and the central perspective relate to the structure of the visual space, that underlies the real extension of the visual perception of space. In his treatise, Berkeley criticized the geometrical-optical methods or approaches in the scientific explanation of the visual perception of space by referring to the unreality or the mere ideality of the geometric lines, as they merely represent the *real* light rays.[11] Berkeley was inclined to a psychophysiological explanatory model for the visual perception of size, position and distance. Johannes Kepler pointed to the falsity of the theories of inference, which assume that the visual size, position and distance perception of objects arise from the inferences of their retinal images.

> Kepler (1604) leaves to the natural philosopher the question of whether the retinal image ›is made to appear before the soul or tribunal of the faculty of vision by a spirit within the cerebral cavities, or the faculty of vision, like a magistrate sent by the soul, goes out from the council chamber of the brain to meet this image in the optic nerves and retina, as if it were descending to a lower court.[12]

Since we cannot see the retinal image, there cannot be any inference between the size of the retinal image and the size of the objects in direct vision.

When Kant declared towards the end of the eighteenth century that space is an exclusively a priori form of sensory perceptions and of the appearances given in sensory perception, he seemed to a certain extent obscure or even suppress the prehistory of the investigation of visual space perception that is briefly outlined above. As previously discussed, the Kantian terms *intuition* (Anschauung) and *appearance* (Erscheinung) relate primarily to visual perception, through which the external objects are mostly given to us, i.e., to the perceiving and cognizing subjects. It is therefore necessary—especially with regard to the discussion of the spatiality of perception and appearance which relate to the spatial expansion of visual space perception—to adequately consider the prehistory of the theories of vision and the science of the geometrical optics, perspectives, etc., that were associated with it. Kant tries to reduce the problem of space in sensory perception contextually to the framework of a *transcendental aesthetic*. The space as a necessary subjective-a priori notion turns out to be a *true* statement; however, this notion veiled many prevailing unsolved problems with regard to the spatial extension of sensibility—especially the spatial extension of vision. In this respect, it can be stated that the programmatic apriorization of space in the context of a *transcendental aesthetic* has nearly achieved a dogmatic status.

In accordance with the transcendental philosophical framework of the Kantian system, the purely empirical terms seem to undergo subtle metamorphoses. The object—or the body—is included in the *Transcendental Aesthetic*, but it is primarily conceived through its effect on the subject in the mode of affecting that brings about

---

[11] Berkeley, George (2009): *An Essay towards a New Theory on Vision*. Philosophical Works, ed. Michael R. Ayers, London, pp. 9–10.

[12] Braunstein, Myron L. (1976): *Depth Perception Through Motion*, New York, p. 5.

a sensation. "Der unbestimmte Gegenstand einer empirischen Anschauung heißt Empfindung.[13]" On a closer inspection, the emphasis in the Kantian transcendental-philosophical system does not lie on the object—as an entity which has a mode of existence completely separate from the subject—but on the "appearance" given in the empirical intuition. For within the framework of his transcendental philosophy, Kant is inclined from the outset to ascribe the object's mode of existence, which is detached from the subject, the status of a *thing-in-itself* (Ding an sich) that eludes the subjective faculty of perception and cognition. Accordingly, the object that in some way affects the human mind is not a mere object, but a *thing in itself*, that is, an unknown "X". The object *exists* either for the subject as *appearance* or in itself—as *thing in itself*. It is well known that the post-Kantian philosophies—especially of Hegel and Schopenhauer—made use of this philosophical position of Kant, on which it was expatiated in their works.

Similar to appearance, perception is more of a transcendental philosophical concept in Kant, under which sensibility—a traditionally empirical concept—seems to be subsumed. Suffice it to say that the objects are given in sensibility (which, as a purely empirical entity, relates to all modes of human sensory faculty) and are conceptually cognized by the mind. According to Kant, just as the empirical intuition, space and time underlie the sensibility (Sinnlichkeit) as pure forms of empirical or sensible intuition. By introducing the concept of perception—between sensibility and understanding—Kant aims at a clear *definition* of pure intuition (reine Anschauung), the a priori forms of which are space and time, and its separation from an empirical intuition. The idea of pure intuition should then philosophically guarantee the apriority of space and time, which according to Kant are not empirical concepts. The purely transcendental status that Kant ascribes to space and time in the context of his *Transcendental Aesthetic* is clearly based on his idea of intuition, i.e., the dichotomy of intuition as "pure" and "empirical." Can a pure intuition—even as a residual entity—simply exist without an empirical *content*? Is it conceivable that a pure intuition completely excludes the direct *sensory experience* of space and time or, more precisely, the spatiality and temporality of objects? Compared to "sensibility", "intuition" seems to acquire the status of a meta-empirical concept in Kant's system of philosophy (which is apparently presupposed by the programmatic *transcendental* metamorphosis of empirical concepts).

Analogies can be drawn between the Cartesian and the Kantian method of systematic separation or negation of subjective attributes and sensory qualities in the object; in both the cases the subjective attributes of the objective qualities, which are mediated solely by sensibility, are methodically negated and separated from the object. The decisive difference between the Cartesian and the Kantian method of negation lies in the ontological determination of the residual entity of space or the spatiality of the object. While the Cartesian method of negation leaves a purely objective extension, as *res extensa*, that is completely detached from the *res cogitans*, Kant conceives the

[13] Kant (1990, p. 64).
See Kant (1998, p. 155, A20; B34). "The undetermined object of an empirical intuition is called **appearance**."

spatiality or the spatial extension of the perceived objects—as an irreducible residual entity—not entirely objective, but rather transcendental—a priori, that is, within the transcendental subject. In the Kantian methodical negation, the extension and shape of the perceived body remain leftover from the empirical intuition; however, "sie gehören zur reinen Anschauung, die a priori, auch ohne einen wirklichen Gegenstand der Sinne oder Empfindung, als eine bloße Form der Sinnlichkeit im Gemüte stattfindet."[14] While in the Cartesian system the residual and, as such, the final entity of the spatial extension of the body is *externalised* as *res extensa* in the perception-theory, in Kant's transcendental system it is *internalized* in a transcendental subject as a pure form of sensory intuition. In a certain sense, a historical-epistemological apriorization of space can be demonstrated between the Cartesian *res extensa* and the Kantian conception of the apriority of space.

The reconciliation between modern rationalism and empiricism in a transcendental philosophical system is considered to be one of Kant's greatest achievements. The divergence between the Cartesian rationalism and the so-called British empiricism was ultimately based on the issue of the origin of ideas, of cognition in the process of knowledge. The apriorism initiated by Descartes and later extensively represented by rationalists like Leibniz was based on the conception that the ideas are a priori, i.e., already present in the subject. In contrast, the empiricists like Locke saw the sources and the initiation of ideas in the a posteriori sensory experiences. Kant tried to overcome this prevailing divergence between rationalism and empiricism in a transcendental philosophical system. A philosophy of synthesis should *combine* the a priori approach of rationalism with the rather a posteriori approach of empiricism. In the transcendental philosophical system, the process of knowledge *begins* a posteriori, i.e. with experience, but both the pure forms of intuition, space and time, as well as the concepts of understanding that form the knowledge, are given a priori. Accordingly, in Kant's system the transcendental-philosophical synthesis forms a certain *application* of the a priori existing pure forms of intuition and concepts of the understanding to the objects (Gegenstände) that are a posteriori given in sensible experiences.

The theoretical foundation of the epistemological synthesis or the binding knowledge, to which Kant's fundamental question "*Wie sind synthetische Urteile a priori möglich?*" (How are synthetic a priori judgements possible?)[15] refers, seems to be determined by a strategic application process in Kant's programmatic reconciliation between rationalism and empiricism. The epistemological synthesis between the a posteriori givenness of objects and the a priori existing forms of intuition and concepts of understanding in the Kantian system of transcendental philosophy is not characterized by a *priority-neutral* reconciliation, but clearly by a preference for the apriority of pure intuition and conceptual knowledge as against the a posteriori givenness of the objects. The epistemological synthesis outlined by Kant—in the mode of an *application* of the a priori pure forms of intuition and concepts of understanding to the phenomena—refers to an original direction of the cognitive process

[14] Kant (1990, p. 64).

[15] Kant (1998, p. 146, B19–20).

from subject to object, which clearly shows the preference of the a priori existence of the forms of intuition and concepts of understanding over the aposteriority of the empirical-sensory experiences and the objects *given* in them.

This preference, on the basis of which Kant—despite his transcendental-philosophical reconciliation between rationalism and empiricism—is ultimately regarded as an important representative of rationalism, seems to be *largely* legitimate within the framework of his philosophy of synthesis. Only within the framework of a transcendental logic can we assume a transcendental application of the concepts of the understanding to the objects given in empirical intuition. For the concepts of the understanding, which are all linguistic concepts, cannot be given as objects in nature and in our empirical intuition; they are a priori constructed by the subject alone (regardless of whether they are already present as a priori constructs or whether their creation in the subject is prompted by the experience—an argument to which Locke seems to be referring in his epistemology). As such, they are necessarily applied to the objects given a posteriori. The "tree" as a pure concept of the understanding is not given either in nature or in our sensory experience; it is to be a priori constructed in the mind and applied in the cognitive process to the a posteriori given appearance of this object, "das der unbestimmte Gegenstand einer empirischen Anschauung ist."[16] The apriority of all concepts and categories of understanding occurs in an unresolvable ontological difference—that is, in a difference in their mode of existence—compared to the a posteriori givenness of the phenomenal objects and therefore necessarily requires an application initiated by the subject in the process of an epistemological synthesis.

The transcendental application of the pure a priori forms of intuition, space and time, to the empirical intuition—or to the *objective* phenomena given in the empirical intuition—seems to be problematic. For within the framework of a *transcendental aesthetic* the apriority of space and time or the a priori presence of the *pure forms* of sensibility in the subject, as Kant conceives of them, does not refer to the conceptuality, but solely to the a priori formality of space and time. But such a formality—especially of space—necessarily relates to a pre-conceptual objectivity. Space as a form necessarily requires a spatial expansion, represented as an objective or merely free-spatial expansion. Particularly in the *irreducible* formality and objectivity of space, both the apriorization of space and the division of this perception-theoretical entity between an a priori existence and an a posteriori givenness appear problematic.

Within the framework of the *Transcendental Aesthetic,* we tried to differentiate an a priori theoretical space of imagination, which forms the basis of all geometrical, and also mechanical and optical cognitions, from an aesthetic space of immediate sensibility. The belongingness of sensory qualities—colour, sound, noise, tactile sensations, etc.—to the subject cannot be doubted. Just like the linguistic concepts, all the sensory qualities of the body, which Locke called secondary qualities, originate solely from the subject and are, as such, a priori representations. In contrast, Locke identifies—following Descartes—the spatial extension of the body and the other attributes that relate to it, such as number, movement, solidity and figure, as primary

[16] Kant (1990, p. 64).

qualities; Detached from the subject, they belong only to the body. The sense qualities arise in the immediate sensibility itself; But they cannot come about as mere qualities (a priori), because they are connected with the spatial expansion of the perceived body in the mode of a synthetic nexus. That is, the spatiality of the sensory qualities is only determined by their bodily and extra-bodily or spatial and objective extension, which, according to Descartes and Locke, has a *real* spatial extension (represented as *res extensa* or a primary quality). From this it can be concluded that both Descartes and Locke's philosophy tacitly presuppose a real objective extension of sensibility. However, Descartes tends towards a fundamentally modal differentiation between the extension of the object perceived purely through senses and its real extension; According to Descartes, the real and the merely mental object form two different modes of objective existence.[17]

For Kant, therefore, the apriorization of space—as a purely a priori form of sensibility—a necessary strategy in order to programmatically consolidate the transcendental trait of the doctrine of aesthetic, which forms the propaedeutic to the entire transcendental philosophy. According to Kant, the spatial extension of sensibility is necessarily an a priori notion. However, space is cognized within the framework of the Kantian *transcendental aesthetic* as an irreducible ontological entity that remains as a final residue after the systematic separation of all subjective attributes and sensory qualities from the object. The subjective impossibility that the spatial extension of the object given in sensibility—compared to all sensory qualities—cannot be negated and *eliminated*, indicates an epistemological or perception-theoretical limit. However, it is rooted more deeply in an ontological finality, namely, in the ontological irreducibility and finality of the space itself—regardless of whether the space is merely a priori represented or actually exists. In addition, the spatial extension of the sensibility relates not only to the objective extension, but also to the free-space, as represented in particular in the visual and auditory perception of extraneous objects. The free-space—as a purely expanded nothing—cannot be differentiated modally between its a priori imaginability and a posteriori givenness. This indistinguishability is clearly based on the ontological irreducibility and finality of the free-space, i.e., of its mode of existence.

While Kant in his emphasis on the apriority of space seems to recognize only the transcendental-subjective *origin* of the notion of space, the necessary existence

[17] See Perler, Dominik (2004): *Inside and Outside the Mind—Cartesian Representations Reconsidered, in: Perception and Reality. From Descartes to the Present*, ed. Ralph Schumacher, Mentis Verlag, Paderborn, pp. 76–77. "There can be two modes of existence for one and the same thing because the form (or structure) of a thing can be present in two different places: in a piece of matter or in the mind. What makes an idea a representation of something is the very fact that the structure of that thing is in the mind. In light of this explanation it does not come as a surprise that Descartes also speaks about the essence of a thing, which is nothing but its basic structure. That it is the same structure that is inside and outside the mind becomes clear from a statement Descartes makes in a letter written in 1645 or 1646: '.. by essence we understand a thing as it is objectively in the intellect, and by existence the same thing as it is outside the intellect …' Here again, it is clear that we are not dealing with two distinct things, but with the same thing that has two kinds of existence. Such a double existence is possible because the essence of a thing can be instantiated both in a piece of matter and in a mind or even in several minds."

of real space is presupposed in his conception of the apodicticity of the geometrical axioms. For especially in the context of *transcendental aesthetic*, the apodicticity of *spatial and temporal structures*—primarily in the area of geometry, and also in other mathematical and scientific disciplines such as mechanics and optics—is based on the perfect conformity between their a priori imaginability and their real givenness or—more precisely—between their a priori and purely objective constructability: "Auf die Notwendigkeit a priori gründet sich die apodiktische Gewissheit aller geometrischen Grundsätze, und die Möglichkeit ihrer Konstruktion a priori."[18] Kant justifiably attributes the apodictic certainty of all geometrical axioms to their mere a priori imaginability or conceivability, yet does not explain how this fundamental relationship between the apriority and apodicticity of the geometrical axioms comes about. The a priori conception of the basic geometric forms and structures are, according to Kant, a priori constructions in our productive imagination, as discussed in detail in the "Axiome der Anschauung" (Axioms of Intuition). But the a priori constructions of the geometric forms and structures, from which the geometric principles (such as the axiom of the straight line or that of the three-dimensionality of Euclidean space) are derived and cognized, necessarily presuppose the *background of free-space*. For all geometric forms and structures are free-spatial forms and structures. Likewise, the *real* existence or constructability of geometric forms and structures presupposes the existence of the real free-space. However, there is no *modal* difference between the purely a priori imagined and the actually existing free-space. Accordingly, the construction of the elementary geometric forms and structures in a priori imagination and their construction in *real* space are more or less based on the same process. The modal indistinguishability of the free-space between its a priori imaginability and its real existence is ultimately based on the ontological irreducibility of (Euclidean) free-space, which manifests—as a purely extended *nothing*—an ontic finality. It is not difficult to see that the apodictic certainty of the (purely) a priori imagined geometric forms and principles is based more deeply on this modal indistinguishability of the free-space. The modal identity of the free-space ensures that the *possibility* of a priori construction of elementary geometric forms and structures is identical with the possibility of their real construction (or emergence).

1.3 The Apodicticity of Transcendental-Aesthetic Intuitions

This *ontological* basis of geometric intuitions and their apodicticity has been sufficiently demonstrated in celestial-mechanical and optical intuitions, especially in early modern astronomy and geometrical optics. Particularly in the case of celestial-mechanical intuitions, one has to conceive the dynamic structures in space necessarily in a priori *imaginations*, as these structures are hardly perceived directly from

[18] Kant (1990, p. 67).

See Kant (1998, p. 158, A24–25). "The apodictic certainty of all geometrical principles and the possibility of their a priori construction are grounded in this a priori necessity."

the earth. A fundamental law of celestial mechanics such as the principle of inertia or, more precisely, the principle of inertial motion, which was first introduced by Descartes and later axiomatized by Newton, unfolded into one of the most important premises for the demonstration of other Newtonian laws such as the ellipticity of the planetary orbits and the area law of planetary motion. This law can be imagined a priori in a geometric-dynamic intuition without experiencing it directly under the ideal conditions in space. Both Descartes and Newton seem to have *visualised* the ideal geometric-dynamic structure of inertial motion in an entirely a priori imagination. However, the apodictic certainty of this a priori conception does not solely relate to its applicability in the real universe, but ultimately to the modal identity between the imagined and the real free-space (in the universe), on which both the *construction* of these dynamic structures in the productive imagination as well as their emergence as real phenomena in space are based. Beginning from the free-space basis of all geometrical and mechanical principles and their construction, the Kantian idea of the apodicticity of the geometrical and mechanical principles, which presupposes the *application* of a priori principles to objective reality, can be reversed: the a priori conceivability of geometric principles and their apodicticity are based on the fact that these axiomatic principles *cannot be* materialized or arise *in any other way* in free-space—regardless of its ontic status as merely imagined or objective-phenomenal free-space. In other words: the apodicticity of the geometrical principles is based more deeply on their origin and constructability in a real free-space—than on their purely a priori imaginability. It is ultimately the *modally undifferentiable* phenomenality and regularity of the geometrical forms and structures that presuppose and safeguard the apriority and apodicticity of their *axiomatic* intuitions.

In the case of geometrical and mechanical intuitions, the forms and structures imagined a priori and their laws are only apparently applied to phenomenal reality; in fact, they are purely *thought* or intuitively visualized in free-space (as in the case of pure geometry) or *with* the objects in free-space (as in the case of celestial mechanics). This means that the spatial intuitions in geometry, mechanics and optics methodically do not follow a purely a priori conception or generation of static and dynamic structures and their *application* to phenomenal reality, but an a priori *co-creation* of real spatial structures, which come about in a modally uniform free-space. Rather, it is a matter of a *resonance* between a priori intuited and phenomenal structures, to which the apodictic certainty of the geometrical, mechanical and optical structures can be traced back.[19] Particularly the spatial and spatiotemporal intuitions in the sciences of geometry, mechanics and optics are based on such a resonance. This

---

[19] This principle, which relates to the static and dynamic phenomena in science, art and in many everyday activities (such as sport), is explained by Martin Kemp in the context of his basic concept *structural intuition.* According to Kemp, in structural intuitions the same structures underlie the internal intuitions and the external phenomena, the visible and invisible structures of which are intuitively visualized: "… the 'structures' are both those of inner intuitive processes themselves and those of external features whose structures are being intuited." {Kemp, Martin (2000): *Visualizations. The nature book of art and science*, Oxford University Press, Oxford, p. 1}. Accordingly, the intuitive visualization of the phenomenal structures of outside world does not require an a priori generation or construction and the subsequent application to reality, but rather insinuates a process-related resonance between internal intuitive and external phenomenal structures: "The structures of

epistemological resonance occurs primarily in the process of visualization, which, as a mode of sensibility, characterizes both the a priori intuition and the a posteriori givenness of the phenomenal structures.

The a priori visualization of the geometrical, mechanical and optical structures *in resonance* with their phenomenality also refers methodically to an epistemological analogization between the domain of the a priori imagination and that of the a posteriori phenomenal givenness of these structures and their principles. The mere idea of the inertial motion in space, and also—in a broader sense—the geometrical deductions of Kepler's laws such as those of the ellipticity and the area law of planetary motion, form analogous miniatures of the real phenomena in space (the actual extension of which goes beyond our imagination). The principle of these and similar epistemological analogies is not just based on a *necessary* perception-theoretical conformity between the a priori conception and deduction and the a posteriori givenness of the (geometrical, mechanical and optical) structures and their laws, but rather on an ontological analogy, as represented by the modal indistinguishability of these structures and their free-spatial background. The above-described epistemological resonance underlying the structural intuitions in geometry, mechanics and optics is ultimately based on this principle of ontological analogy. However, this primordial principle of analogy does not just induce a pure application of the a priori conception to the phenomena, but rather enables the subject to identify or even *dictate* the conformity between the a priori intuition and the a posteriori givenness of phenomenal structures.

The transcendental-aesthetic intuition owes its apodictic certainty—with regard to its spatial and spatiotemporal structures—clearly to this principle of epistemological analogy, which is based on the resonance between the intuitive and the phenomenal structures. But intuition is basically a process; the processuality of intuition in its apodicticity points to an equally apodictic certainty, which surfaces particularly in the history of celestial mechanics in various traits—as convictions, predictions, in addition also as aporias. The a priori notion of inertial motion and the centripetal gravitational attraction of the sun and all celestial bodies as well as the (Newtonian) laws deduced from these and other premises, such as ellipticity and the area law of planetary motion, do not arise solely as knowledge (a priori); their epistemic status is inevitably intertwined with the conviction that these celestial mechanical structures, represented in analogies, should correspond to the real and immeasurably extensive dynamic structures in space. The three-dimensional models of the

---

the external world within which we need to operate […] are those with which the internal structure of intuition has been designed to resonate, continuously reinforcing and retuning themselves in a ceaseless dialogue of matching and making" (ibid., p. 1–2). Especially in the concept of structural intuition, the epistemological dichotomy between the purely a priori conceivability and the a posteriori givenness of phenomenal structures seems to be suspended—as manifested in sciences such as mechanics and optics, in plastic art and in many everyday events. The resonance between intuitive and phenomenal structures (in Kemp's conception of structural intuition) can also be traced back to the above-discussed modal indistinguishability of free space, in which the structures arise or are constructed independently of their a priori and a posterior status, and also to their undifferentiated ideality as spatial structures.

celestial mechanical structures, which are a priori imagined or visualized, and their graphic representations are basically miniatures that *correspond* to the real dynamic structures in space and thus enable us to discover their laws in our *earthly* intuition and deduction. The basis of this process, on which the veracity and apodicticity of the a priori celestial mechanical knowledge build up, is obviously the modal identity and indistinguishability of the geometrical-mechanical forms and structures and that of the free-space in which they arise, as has been discussed above. This modal identity, which constitutes the principle of the epistemological analogy between the imagination or deductive representation and the *reality* of the celestial-mechanical structures, can ultimately be founded on the ontological irreducibility and the finality of these structures and the free-space—both in their imaginary conception and in their phenomenal existence.

When the understanding of a phenomenon comes up against a limit, an aporia arises—the state of puzzlement and helplessness in the process of thought and knowledge. The aporia in thinking and cognition usually leads to subjective speculations of the phenomenal reality, the basic doxastic trait of which ultimately owes its legitimacy to the unprovability of the phenomenon. Numerous examples of phenomenal and subjective aporias can be identified in the history of philosophy and natural sciences, such as the existence of the infinitesimals, the causality or the *mechanical* mode of action of gravity at distance, bodily and extra-bodily extension of sensibility—especially the emergence of visual space perception—the mode of the existence of space, the divisibility of the body, etc. The *fiction* of the aether in free-space, as assumed by philosophers and scientists, was basically a speculation that gave rise to some of the aporias of natural phenomena, such as the material mediality or the mechanical causation of gravity or the movement and dispersion of light in free-space. This also includes the conceptions of corpuscles and atoms as irreducible and as such final components of matter and of the forms ascribed to them, which should *cause* the material and subjective-sensory qualities of the phenomenon. The phenomenal aporia thus seems to give rise to a variety of doxastic speculations, the certainty and veracity of which we are unsure of or puzzled. However, the phenomenal aporia can lead to the epistemological apodicticity of the phenomenal reality, as the *subjective* puzzlement in the cognitive process *presupposes or rather necessitates the assumption and unconditional acceptance of the phenomenality.* That is, the phenomenal aporia would induce the subject to apodictically (not doxastically) accept an unrecognized phenomenal causality. Thus the aporia can lead to apodicticity of knowledge, the objectivity of which is yet to be proven.

The analogicity of intuition between a priori imagination or representation and a posteriori givenness of phenomena enables the intuition to a priori secure the objective phenomenality. Even when intuition encounters an aporia in this process, it can develop into the apodictic certainty and subsequently into the conviction *that there is an unexplored or undiscovered phenomenon in reality that can resolve this aporia, and for which the empirical and experimental science should provide evidence.* The aporia, in the synthetic unity of intuition, can lead to the apodictic certainty of knowledge, that we do not acquire immediately, but can *sense* it and become aware of its veracity. Accordingly, the synthetic unity of the sensory or transcendental-aesthetic intuition can enable an aporia to unfold into an apodictic knowledge. The investigation of sensibility has two important objects, namely the sensory qualities, which are perceived subjectively, and the spatial and temporal structures, over which the senses extend. Accordingly, the intuitive investigation of the spatial and temporal extension of sensibility forms a *structural intuition* of sensibility. It is true that the absolute (ontic) differences of sensory qualities prevent the scientific conception of their mutual inferences or suggestions, but the ontic unity of space and time and the modal indistinguishability between the apriority and the a posteriori givenness of these forms of sensibility enable us to draw analogies between the extensions, i.e., between the spatial and temporal *embodiments* of sensory qualities. The investigation will therefore begin with the question of whether the directly experienced extension of sensibility is based on a real or phenomenally given spatiotemporal structure. Sensibility is thereby detached from the prevalent notion of its subjective apriority by tracing and ascertaining the *objective* participation, i.e., how and to what extent the perceived objects *participate* in the sensibility.

References

Berkeley, George (2009). *An Essay towards a New Theory on Vision.* Philosophical Works. Michael R. Ayers (Ed.). London.

Braunstein, Myron L. (1976). *Depth Perception Through Motion.* New York.

Descartes, René (2015). *Die Welt.* Christian Wohlers (Trans. & Ed.). Meiner Verlag. Hamburg.

Kant, Immanuel (1990). *Kritik der reinen Vernunft.* Raymund Schmidt (Ed.). Meiner Verlag. Hamburg.

Kant, Immanuel (1998): *Critique of Pure Reason.* trans. & ed. Paul Guyer and Allen W. Wood. Cambridge University Press. Cambridge.

Kemp, Martin (2000). *Visualizations. The nature book of art and science.* Oxford University Press. Oxford.

Morgan, Michael J. (1977). *Molyneux's Question. Vision, Touch and the Philosophy of Perception.* Cambridge University Press. Cambridge.

Perler, Dominik (2004). *Inside and Outside the Mind—Cartesian Representations Reconsidered.* In: *Perception and Reality. From Descartes to the Present.* Ralph Schumacher (Ed.). Mentis Verlag. Paderborn.

Chapter 2
The Analogicity of Senses

2.1 The Spatiality of Senses

Our sensory perceptions, where we directly experience the spatial extension and localization of the various sensations, are not identical, but they are analogous to one another. While the tactile sensations, pain or taste are bodily extended and localized, our visual and auditory sensations have extra-bodily extension and localization. The sense of smell is largely localized in the body, i.e. in the nose, but it enables us to vaguely perceive its external localization in objects. Since the sensory perceptions turn out to be completely different in terms of their modalities, their analogicity is based—in addition to their subjective origin—on their spatiality, i.e. on their spatial extension and localization.

Only through our senses are we able to perceive the extension of our own body and that of the world in which we, as living beings, find ourselves among all other objects. All bodily and extra-bodily sensations enable us to experience the spatial extension of our body and the surrounding world. Although we are aware of the extension of our body and the world (by trusting our senses, which convey this elementary knowledge), we tend to be reluctant to ascribe to sensibility a *real* or primary quality of spatial expansion. For, in a strictly Cartesian sense, we are inclined to differentiate ontologically and modally the existence of our subjective sensory perceptions from that of our purely objective body and the objects in the world, and segregate them subsequently from one another. Accordingly, the *apparent* extension of all bodily and extra-bodily sensory perceptions is assigned a different ontic status—that is, a non-physical mode of being.

The subjectification or the subjective apriorization of sensory qualities turns out to be historical in the context of modern philosophy. The complete reduction of the sensory qualities to purely mental phenomena in the Cartesian system is, however, preceded by a transitional phase between the late Middle Ages and the early modern period. In late medieval scholasticism, the mental origin of sensory qualities was acknowledged, but their localization and extension *in objects* were not negated.

B. Thaliath, *The Embodiment of Senses*, https://doi.org/10.1007/978-981-96-8468-7_2

Colours are subjective sensations, but they are perceived in the object or in the *real* objective extension.[1] This originally *aesthetic* nexus—between a perceiving subject and the perceived object—clearly indicates that the object is *directly involved* in sensory perception. The *real* in this objective participation is obviously constructed and safeguarded by the spatiality of the sensory perception, i.e. by a primary quality. In the case of external visual and auditory perceptions, this spatiality or the spatial basic structure of sensibility shows itself not only in an objective localization, but also in a free-spatial distance, extension and organization. Direct vision is based on a perspectival spatial structure that includes both the free-spatial and objective extension.

While the late medieval scholasticism epistemologically recognized the corporeality and the extra-corporeal objectivity of sensibility, Descartes, who initiated the historical transition from the Middle Ages to the modern age in his philosophy, pleaded for a complete separation of the sensory qualities and their *reality* from the phenomenal world of the body and the extraneous objects. To this end, Descartes developed a certain philosophical method, namely the systematic negation and separation of all mental attributes and sensory qualities from the object, the real *existence* of which can be doubted.[2] What remains after the Cartesian method of doubting and systematic separation are two absolutely final and irreducible entities, namely the *res cogitans*, to which all the mental attributes and sensory qualities negated and separated from the object were ascribed, and the *res extensa*, i.e. the pure spatial extension, cognized as the only objective mode of existence that cannot be negated. According to Descartes, the spatial extension of objects (including the bodily extension) is the sole factum of phenomenal reality that is not contingent upon the *thinking* subject and therefore has an *ontological status* completely autonomous from the *res cogitans*.

However, the method of negation prompted a decisive polemic against the Cartesian differentiation between *res cogitans* and *res extensa* in Descartes' time and in the post-Cartesian philosophy of the modern age. This polemic was expressed in the

---

[1] In her seminal work, *Zwei Untersuchungen zur nachscholastischen Philosophie*, Anneliese Maier refers to the late-medieval-scholastic assumption of the *real* extension of sensory qualia, which was systematically negated in Cartesian modernism: „für die Scholastik entstehen die qualitates secundae aus den primae im Objekt und nicht erst, wie für die späteren, im wahrnehmenden Subjekt. Ihre Realität wurde darum in der traditionellen Philosophie nie in Zweifel gezogen, und ebenso wenig die Abbildlichkeit der Qualitätsempfindungen. [...] Wie die Qualitäten im einzelnen von den primären abhängen sollen, wird, besonders wenn es sich auch um die nicht-taktilen handelt, in der älteren Philosophie nur sehr undeutlich gewusst und gesagt. Die Argumentation geht häufig über die Vorzugsstellung des Tastsinns, denn der ist zwar nicht der vornehmste, aber der notwendigste Sinn, der von allen vorausgesetzt wird, selbst aber keinen voraussetzt. Die Betrachtung wird damit auf ein Gebiet hinübergespielt, das vielleicht die stärkste Problematik und die meisten Ansatzmöglichkeiten für die Weiterentwicklung enthielt." See Maier, Anneliese (1968): *Zwei Untersuchungen zur nachscholastischen Philosophie*, Rome, p. 18.

[2] The method of doubting and systematic separation was introduced and clearly explained in the famous parable of wax in the second book of meditations. See Descartes, René (2009): *Meditationen*, trans. Christian Wohlers, Meiner Verlag, Hamburg 2009, pp. 33–35.

question of Princess Elisabeth of Böhmen and Pierre Gassendi about the immateriality and the *unextendedness* of the soul, which nevertheless causes physical acts of volition, as well as in the curious plea of John Locke for the recognition of the solidity of the object—in addition to the Cartesian res extensa—as a mode of existence of phenomenal reality that cannot be subjectively negated and segregated. Before we discuss the polemics of Princess Elisabeth and Gassendi in the context of our investigation, we try to address Locke's position, in which the Cartesian reduction is largely supported but, at the same time, decisively revised. Solidity is a necessary factum that characterizes the object's mode of existence, which is completely separate from the subject and is autonomous. Solidity is also an inseparable trait of materiality, which gives phenomenal reality to the extension of objects. If the *res extensa* points to an irreducible and not to be negated *material extension*, this objective materiality is not only characterized by the mere spatiality, but necessarily by the material solidity and should, as such, be part of the mere spatial extension. In other word, the material solidity should be included to the category of primary qualities that cannot be negated *in* the body. The emphasis on the solidity of the body as a primary quality relates primarily to the materiality of the body, which alone defines the spatial expansion of the body as *res extensa*, and which therefore impart the physical bodies of their autonomous existence from the subject. Locke's addition of solidity to the primary qualities basically extends and revises the Cartesian differentiation between *res cogitans* and *res extensa*. If the purely spatial extension—the *res extensa*—is the sole residual factum that remains in the object from the subjective negation and separation of all mental attributes and sensory qualities, the object can no longer *exist*. For the purely spatial expansion dissolves into the *free-space*, which is merely an extended nothing. On closer inspection, the spatial extension and demarcation of the object not only presupposes a purely spatial extension—as *res extensa*—but also the objective solidity. The absolutely empty spatiality of the object cannot draw any boundaries against the surrounding free-space (and thereby guarantee the existence of the object); for this the *res extensa* needs the inclusion of materiality, the essential trait of which is inevitably solidity. In short: the Cartesian *res extensa* should necessarily *house* the objective solidity, so that the existence of the object—phenomenal reality in general—is safeguarded. However, the Lockean revision seems to be rather insignificant in the *context* of Cartesian philosophy. After all, Descartes understands spatial expansion as a material extension, as space is primarily defined by the body. According to Descartes, space and material bodies are inseparable entities; the *res extensa* originally claims the materiality of the extended body. The mere emptiness, i.e. the absolutely empty spatial extension, cannot exist. Locke, on the other hand, advocates an atomistic doctrine where material atoms are ontologically differentiated from empty free-spaces.

In addition, every form of subjective sensibility, in which a corporeal or extracorporeal object is perceived, necessarily presupposes a material causality. When we perceive an object in colour, this subjective-sensory perception is not only caused by the spatial extension, but also by the materiality of the object. If we, like Descartes, negate our subjective sensation and try to separate it from the object, we only negate the effect of the object on us and not the objective-material cause of this effect. Hence every negation of the sensory perception, which is basically an objective effect on our sensory faculties, presupposes its material causality in the object, which cannot be negated. Now a domanial difference between the purely mental reality and its material causality should be observed. Strictly speaking, the Cartesian negation is limited to the domain of the subject that alone can construct the reality of sensibility. The material causality of sensible reality belongs to the domain of phenomena that is completely different from the subject. As previously discussed, in the Cartesian system, the method of doubt and the negation and separation that follow it produce two residual and absolutely irreducible modes of existence in the world—as *res cogitans* and *res extensa*. However, following Locke, if we add the purely material solidity as an irreducible factum to the *res extensa* that remains in object as a residual entity, the Cartesian method of reduction seems to experience a setback from the outset. That is, we can hardly add the materiality of the phenomenal world to the Cartesian residual entities—the unextended and immaterial mind and the pure extension of the body. Such a critical addition would signal a failure of the absolute (Cartesian) reduction of physical existence to a *res extensa*. For the materiality would *fill* afresh the empty *res extensa* and thereby rehabilitate all the material causalities of the subjective-sensory realities in the domain of the phenomenal world that Descartes tacitly neglected and negated.

The irreducible materiality will then not only underlie an original causality of sensibility, but also the mediation—in the mode of a chain of material-causal nexus—between the perceived object and the brain. The causal basis of a sensory perception is not limited to its *origin* in an object, but extends to the entire material and physical processes in and outside the body, which connects the perceived object with the brain that processes the *material* content of perception and develops it into a sensory perception. A pain or cold sensation that is localized in the foot has an originally bodily or extra-bodily material cause—such as a physiological inflammation or the coldness of the snow—but the reality of the sensation of cold or pain in the foot is necessarily *caused* by the neural mediation in the nervous system, which connects the foot, i.e. the part of the body where the sensation is located, with the brain by means of electrical phenomenon and the extremely simultaneous transmission of its signals. In *external* seeing and hearing, the material-causal mediation of sensory perception extends to external phenomena such as light or acoustic waves. The perceived colour of an extra-bodily object results from a chain of optical and other physical, i.e. photoelectric, chemical as well as neural causes—in addition to an original causation, namely the *material* properties of its surface, which reflects a certain wavelength of the light falling on the object. All these causes belong to the domains of matter, namely the geometrical-optical formation of the retinal image, the photoelectric processes on the retina and the neural transmission of the photoelectric input from the retina into the brain cells where it is processed and is developed into a visual sensation.

In the context of sensibility, a purely spatial extension of the object from which all sensory qualities are separated can hardly be imagined. The *res extensa*, as an absolutely residual entity, is not a mere emptiness, but it has invariably a material fullness and solidity.[3] As the materiality of phenomenal reality brings about both the original and the medial causality of sensory perceptions, the different modes of senses seem to be subject to a *real spatial structure*. When Descartes recognizes *res extensa* as the absolutely final and irreducible mode of existence of the body, which cannot be subjectively negated and separated, he also legitimizes the real or objective existence of space—but not as a mere carrier (according to the Aristotelian idea) but as the inseparable mode of existence of the body itself. Such an objectivity or *phenomenality* of space is hardly recognized in Kant's transcendental philosophy, that is, in a later system of modern philosophy. According to Kant, space and time are necessarily a priori representations; the spatiality and temporality or the spatial and temporal *existence* of a object perceived through senses is *primarily* a priori determined by a transcendental subject.

In the Cartesian method of doubt and negation, the sensory qualities are withdrawn from the perceived object as purely a priori ideas, nevertheless the final and irreducible entity that remains of the object is an objectively given mode of existence of the object—a *res extensa*. This means that in their epistemological dualism Descartes—and more clearly Locke—act on the assumption that the mental and the phenomenal domains are absolutely different from one another. In his doctrine of transcendental aesthetic, which is the first part of the transcendental elementary doctrine (*Transzendentale Elementarlehre*) and the propaedeutic to his transcendental philosophical system, Kant more or less tries out the method of systematic negation that has been handed down in modern philosophy. However, Kant proceeds solely from the *standpoint* of a transcendental subject, where he does not ascribe to the forms of sensibility, namely space and time, an objective mode of existence, but rather a transcendental-subjective apriority:

> Wenn ich von der Vorstellung eines Körpers das, was der Verstand davon denkt, als Substanz, Kraft, Teilbarkeit usw., imgleichen, was davon zur Empfindung gehört, als Undurchdringlichkeit, Härte, Farbe usw. absondere, so bleibt mir aus dieser empirischen Anschauung noch etwas übrig, nämlich Ausdehnung und Gestalt. Diese gehören zur reinen Anschauung, die a priori, auch ohne einen wirklichen Gegenstand der Sinne oder Empfindung, als eine bloße Form der Sinnlichkeit im Gemüte stattfindet.[4]

The apriority of space and time—as pure forms of sensibility which can exist in mind "even without a real object of the senses or sensation"—forms a basic assumption in the Kantian system, which, however, can be legitimized only in the context of Kant's transcendental philosophy.

---

[3] Here the property of solidity—as it was presented by Locke (as a primary quality)—next to liquid or gaseous form does not refer to a mere mechanical state of the material body, but it encompasses all these modes of material existence and characterizes the irreducible trait of the material body itself.

[4] Kant (1990, p. 64). See Kant (1998, p. 173, B35, A21). "So if I separate from the representation of a body that which the understanding thinks about it, such as substance, force, divisibility, etc., as well as that which belongs to sensation, such as impenetrability, hardness, color, etc., something from this empirical intuition is still left for me, namely extension and form. These belong to the pure intuition, which occurs a priori, even without an actual object of the senses or sensation, as a mere form of sensibility in the mind."

Within the framework of modern philosophy, i.e. from the Cartesian assumption of a *res extensa* to the Kantian ascertainment of the perfect apriority of spatial intuition, a historical apriorization of space is identifiable. In the Kantian *Transcendental Aesthetic*, the objective extension of sensibility is ascribed from the outset to a transcendental subject. Space and time are at the same time forms of intuition as well as appearance (Erscheinung) that is given in the empirical intuition; as pure forms of sensibility, they are a priori present—i.e. before all experiences—in the (subjective) mind. Space and time as pure forms of a priori sensory intuitions form a common and irreversible basis in the Kantian system for all modes of sensory perception, which we can justifiably call the a priori structure of sensibility. I.e. the *Transcendental Aesthetic* programmatically isolates the spatial and temporal structures of sensibility and declares them as purely a priori forms of intuition and appearance. Being irreducible and final forms of existence, these structures can neither be withdrawn from empirical intuition nor from appearance. Corresponding to the unity of space and time, which underlies all modes of sensibility, Kant uses terms such as *Anschauung* and *Erscheinung*, which refer to sensibility in general (although both terms primarily indicate seeing).

2.2 The Analogicity of Sensory Structures

In the Kantian doctrine of *Transcendental Aesthetic*, space and time as purely a priori forms or structures of sensibility manifestly refer to the spatial and temporal *extension* of sensibility, represented at the same time as the extension of the empirical intuition and the extension of the appearance given in it. Through this doctrine, Kant creates a general (a priori) basis for all types of sensibility, which in their qualitative existence prove to be completely different from one another. As sensory qualities alone, the visual, auditory, tactile, taste and olfactory sensations have nothing in common; they are absolutely different states and processes of sensibility. There are clear qualitative differences between these fundamental modes of sensibility, which constitute the modal and ontic differentiation and autonomy of sensory perceptions. Accordingly, different sensory perceptions cannot be analogized as mental-qualitative states. However, they have a common basis in their a priori formality, i.e. in their a priori spatial and temporal extension. Although the spatial extension of vision modally differs from that of touch and that in turn differs from the extension of other sensory perceptions, the unity and irreducibility of space and time impart the various modes of sensibility an analogous basic structure.

It is precisely on the spatial and temporal extension of sensibility that the analogicity of sensory perceptions is based. The spatial and temporal extensions of the various sensory perceptions are not identical, but analogous to one another. The visual perception extends in a perspectival spatial structure, whereby the spatial extension of the auditory perception manifests itself as a purely linear distance of the object heard (and not as a perspectival convergence, as happens during seeing).

Compared to the external and geometrical-perspectival structure of vision, the physical extension of touch or taste does not turn out to be geometrical. The analogy of sensibility or the extension of sensory perceptions therefore does not enable modal analogization, but is based solely on the ontic unity, i.e. on the irreducibility and finality of the forms of sensibility, namely on space and time. Some attributes of the spatial and temporal extension of sensory perception show a clear analogy in their modality, e.g. the spatial localization of senses. The sensory perceptions are localized in the body and in the external objects. Just like the pain in the foot or the sweetness on the tongue, the colour or tone is felt or sensed localized in an object. In principle, the subject perceives the localization of different sensory perceptions as merely spatial—bodily and extra-bodily—distances that can be analogized in a *uniform* mode of spatiality.

As discussed above, in both the Cartesian and the Kantian method of systematical negation of sensory qualities in the object and in their subsequent apriorization, reference is made solely to their *subjective* reality and not to the *material* causality on which this reality is based. In the Cartesian and Kantian systems, sensibility is *modally* differentiated and separated from the domain of material objects; however, the reality or the ontic state of sensibility is inevitably based on an *ontological* material-phenomenal causality. However, in Kant and Descartes too, the materiality of the body was *tacitly* viewed as a composition of all subjective mental attributes and sensory qualities—in the context of their epistemological method of negation and separation—and consequently negated in the object, so that the mere spatial extension of the object, which cannot be cleared out or *emptied*, remains.[5] But, following Locke, if we recognize the materiality of the object—as well as the spatiality—as a mode of existence that is completely independent of the subject and, as such, cannot be negated, and if we subsequently rehabilitate it in the Cartesian and Kantian discourse on sensibility, we ascribe to the reality of sensibility a material causality that can neither be negated nor separated from the modes of existence of sensibility. The epistemological rehabilitation of materiality and its addition to the residual entities left over (primarily to the *res extensa*) after the Cartesian and Kantian methods of negation will lead in this way to a decisive setback of this method. For henceforth one does not see the reality of sensibility in its bare unity of effectuation and otherness (from the material object), but necessarily in their rootedness in a material causal basis— that is, in an ontological-etiological totality.

---

[5] The Cartesian idea of *res extensa* indeed points to a *material* extension, but Descartes hardly emphasizes the materiality of the physical extension, which is left over from the negation and separation of all mental attributes and sensory qualities. In the parable of wax (in *Meditations*) Descartes tends to reduce the essential trait of the residual *res extensa*, namely the flexibility of the form of extension, to a unified idea of spatial extension. But the flexibility of the form (of the wax from which all sensory qualities disappear) is a *material* flexibility and as such presupposes an irreducible and residual materiality. When Locke adds the property of solidity to the Cartesian notion of material *res extensa* (which differs from the merely mental *res cogitans*), he seems to legitimise the problematic reduction of objective or bodily existence to a mere *res extensa*. For the Cartesian conception of *res extensa* implies the materiality of the object inevitably as a material extension. But this implication remains invalid as long as it is not emphasized.

The polemic of Princess Elisabeth of Böhmen and Pierre Gassendi against the Cartesian reduction of the soul to a *res cogitans*, or to an immaterial substance without extension, points to the need to rehabilitate extension and materiality in the domain of subjective acts of volition and sensibility. In her first letter to Descartes, Princess Elizabeth of Böhmen asks how an immaterial and unextended soul can influence a material and extended body and induce it to the acts of volition. According to Princess Elisabeth, the mechanical acts of volition, shown above all in the bodily movements, necessarily presuppose the extension and materiality of the soul. As is well known, Descartes could hardly ignore Princess' question. In his answer to her question, Descartes admitted that, as far as the acts of volition and sensory perceptions were concerned, he was silent about the indispensable connection between body and soul in favor of thinking and its complete otherness and separation from the body:

> Es gibt zwei Dinge in der menschlichen Seele, von denen die ganze Kenntnis, die wir von ihrer Natur haben können, abhängt: das eine ist, dass sie denkt, das andere, dass sie, da sie mit einem Körper vereinigt ist, mit diesem handeln und leiden kann. Ich habe fast nichts zu diesem Letzteren gesagt und mich nur bemüht, das Erste gut verständlich zu machen, weil es meine hauptsächlichste Absicht war, den Unterschied zwischen Seele und Körper zu beweisen. Dazu konnte nur dieses dienen und das andere wäre hier schädlich gewesen.[6]

However, Descartes presented all forms of sensibility as modes of thinking in *Meditations*.[7] This means that Descartes sought to subsume all subjective operations in perceptual, cognitive and thought processes under a generic term, more precisely a generic context of thinking, and tried to ascribe neither extension nor materiality to it—as compared to the world of merely extended bodies. Descartes tried to enlighten the Princess about the mysterious connection between body and soul, which is expressed in the acts of volition and sensory perceptions, by means of an analogy with regard to the connection between the force of gravity felt subjectively and the body that gravitates towards earth.[8] The princess, however, seemed hardly satisfied by Descartes' explanations; As before, she remained firm on her conviction that the soul without extension and materiality is unconceivable:

> Wie die Seele (die unausgedehnt und immateriell ist) den Körper bewegen kann ... Und ich gestehe, daß es mir leichter fallen würde, der Seele eine Materie und eine Ausdehnung zuzuschreiben, als einem immateriellen Wesen die Fähigkeit, einen Körper zu bewegen und von ihm bewegt zu werden. Denn, wenn das Erstere durch Information geschähe, müßten die Lebensgeister, die die Bewegung verursachen, intelligent sein, was Sie aber keinem körperlichen Wesen zubilligen. Und obwohl Sie in Ihren metaphysischen Meditationen die Möglichkeit des Zweiten nachweisen, bleibt es gleichwohl sehr schwer zu verstehen, wie eine Seele, so wie Sie sie beschrieben haben, nachdem sie die Fähigkeit und die Gewohnheit des folgerichtigen Denkens erworben hat, dies alles durch einige Dünste verlieren kann und

---

[6] See Ebbersmeyer, Sabrina (ed. 2015): *Der Briefwechsel zwischen Elisabeth von der Pfalz und René Descartes*, Wilhelm Fink Verlag, Paderborn, p. 29.

[7] Descartes, René (1972): *Meditationen*, trans. and ed. Artur Buchenau, Felix Meiner Verlag, Hamburg, p. 145.

[8] Descartes, René (2003): *Meditations and Other Metaphysical Writings*, trans. Desmond M. Clarke, Penguin Classics, London, p. 150.

wie sie, die doch ohne Körper existieren kann und mit ihm nichts gemeinsam hat, dermaßen von ihm beherrscht werden kann.[9]

The Princess' polemic mainly relates to the acts of volition of the body, though it also applies to the receptive sensory perceptions, as can also be found in Descartes' replies. While the spatial extension of sensibility is experienced directly, its materiality seems to be a problematic assumption. For the existence of the purely mental modes of sensibility is completely different from the existence of the material body—a distinction on which the Cartesian method of systematic negation and separation of sensory qualities is based. In the mode of existence the sensibility is completely different from the material objects; the nexus between these absolutely different *ontic* domains is essentially not modal but *causal*. The *reality* of sensibility is necessarily based on a material causality. Accordingly, an *ontic synthesis* between subjective sensibility and phenomenal reality seems impossible. For the ontic emergence of sensibility from the phenomenal materiality turns out to be a complete emergence—as compared to the ontic emergences in the domain of material phenomena. In reality, no residue of a phenomenal materiality remains from the *qualitative* emergence of sensibility—of seeing, touching, hearing, tasting and smelling as well as other bodily sensations.

The recognition of a *complete emergence* of sensibility, however, does not exclude its material-phenomenal causal basis; instead, it even presupposes the immanent etiological structure that *causally* connects the existence-sphere of sensibility with that of the material phenomenality. The material-causal basis of sensibility offers—in addition to the spatial extension—a platform for the analogy of the various modes of sensibility. Although different sensory perceptions arise from equally different physical-material processes, the causal processes do not differ entirely from one another. There are certain phenomena in these physical-material causal processes, which all sensory perceptions have in common and that give rise to their analogous characteristics, such as the spatial extension and localization. The neural processes that ultimately cause the various modes of sensibility turn out to be completely different from one another in their modalities; nevertheless they are based on identical phenomena. An appropriate example is the electrical phenomenon that underlies all processes in the nervous system, namely the transmission, processing and the (bodily) spread of sensory perceptions. This also includes chemical and biological processes. These *unified* physical-material phenomena appear to accomplish some analogous characteristics of the various modes of sensibility, namely the bodily and extra-bodily extension and localization of senses.

We are able to directly experience the extension and localization of our bodily sensations, such as pain, coldness, all tactile sensations, taste, etc. Just as we are aware of the spatially extended mode of existence of our own body, we perceive the bodily extension and localization of our sensations in their reality or in their real existence. The reality or real existence of the bodily sensations is secured and guaranteed here by the direct participation of the object or the body in sensibility; the absence of the objective body and its participation leads to the virtuality of bodily

[9] Lauth, Bernard (2006): *Descartes im Rückspiegel*, Paderborn, p. 189.

sensory perceptions, as represented particularly in the phenomenon of phantom pain or phantom limb. The involvement of the body, through which the characteristics of bodily sensations such as extension, localization and also the simultaneity of sensations come about, relates to various material, i.e. physical-physiological phenomena and processes. The neurons in the brain and in the entire nervous system, which process the bodily sensations and spread or localize them in the body, consist of electrical and biochemical phenomena and processes.

The senses of sight and hearing extend outwardly in the open space to external objects. What could be the nature of the material-bodily participation that would then bring about and guarantee the reality of the external extension and objective localization of the sense of sight and hearing? Would it be possible to draw analogies between the bodily and extra-bodily extension, localization and simultaneity of the sensations—in view of the generality of the above-discussed material-bodily participation? If in the reality of our external visual and auditory senses—in contrast to our bodily sensations—we do not recognize any external-material participation and mediation of a *real* phenomenon, we reduce our visual and auditory perceptions and their directly experienced characteristics to mere virtualities—to phantom-images, -voices or -sounds. Our experience contradicts such a reduction from the outset; we perceive the colour and tone of an external object localized on the object, just as we perceive the sensation of pain, taste, warmth or cold to be extended and localized in our body. But if we recognize the analogy between the bodily and extra-bodily sensory perceptions, especially with regard to their spatial extension and localization as well as their temporal simultaneity, we tacitly assume that the corporeality of our sensibility, which we can experience directly, also extends extra-bodily *in reality*, i.e., into the immeasurable distance and spatiality of vision and auditory perception. That is to say, just as we can localize the pain in our body or bodily touch and feel an external object and thereby experience a certain tactile sensation such as cold, warmth or softness, we *touch* the external and spatially distant objects *optically* and *aurally* and feel their colour and brightness as well as their tone, noise, or voice. Such an assumption will elude the common human imagination.

How can an adequate analogy be drawn between bodily and extra-bodily sensations? The philosophical standpoints alone, such as Kant's transcendental reductionism in his *Transcendental Aesthetic*, seem to be inadequate here; they do not necessarily tend towards an epistemological finality, but rather towards a doxastic contextuality. If, on the other hand, the assumption is made solely from the givenness of the phenomena, regardless of the subjective standpoints from which they are viewed, the analogization could methodically—within the framework of the natural sciences—lead to the discovery of phenomena which should necessarily be given objectively. The phenomenal-causal basis of all sensations discussed above is given in the body. The neural processes in the body that cause sensations are also given and as such cannot be determined or controlled by the sensing subject. The bodily and extra-bodily extension, localization and simultaneity of sensibility are by nature mental states, but they are based on an objectively given phenomenal-causal structure. This structure has some traits that all modes of sensation have in common or which, in an undifferentiated manner, give rise to the above-mentioned properties of all sensory perceptions. One of them is the electrical phenomenon in the neural processes in the nervous system and in the brain, which not only brings about the bodily extension and localization, but also the simultaneity of bodily sensations. The simultaneity of bodily sensory perceptions provides adequate evidence that the physical localization of the sensory perception necessarily arises from electrical signals in the nervous system. When we touch a cold or warm object with our foot or with the fingers of our hand, we immediately feel the warmth or cold localized in the foot or in the fingertips. This sensation process includes the following stages: (1) the physical touch of the object, (2) the transmission of this sensory input to the brain through the nerves, (3) the neuronal processing of the transmitted input in a certain part of the brain and (4) finally, the sensation localized in the body, i.e. in the foot or in the fingers. From the first stage of physical contact to the final result, namely the subjective sensation, this process of sensory perception occurs in extreme simultaneity. The simultaneity and bodily localization of the sensation are obviously based on the simultaneity of the transmission and processing of the signals in the nervous system, which should primarily be an electrical phenomenon; a chemical or fluid dynamic phenomenon or process alone cannot facilitate the *temporal* simultaneity of bodily sensations.

These insights should prove to be a priori, even if we are not individually aware of the neural network in our nervous system and its functions—especially the transmission of the raw sensory input to the brain and that of the sensory perception back to the place or *spot* of sensation in the body. Because the simultaneity of our bodily sensations—e. g. that we feel the cold or warmth of an object immediately or at the same moment in our fingers when we touch it with hands—can only arise through the simultaneity of the electrical phenomenon. A chemical or fluid dynamic phenomenon (such as the blood circulation in the body) can never achieve the speed and simultaneity of the electrical signals, especially when this simultaneity applies not solely to a one-way, but to a two-way transmission—i.e. from one place in the body to the brain and back. Neurobiological research has shown that it is the flow of electrochemical signals between sensory neurons in our nervous system that brings about the simultaneity of our bodily sensations and all simultaneous acts of volition.

The electrical signals in the nervous system also ensure the physical extension and localization of (bodily) sensory perceptions that are processed in the brain. The fact that we simultaneously feel pain, all tactile sensations, taste, etc., localized in certain places or spots in the body, is thanks to the electrical signal system in the body, which is centered in the brain and is present throughout the body through the nervous system. All sensory perceptions are purely subjective in nature, but their bodily extension, localization and simultaneity are inevitably based on a phenomenal causality of electrical or electrochemical processes.

However, electricity or the electrical signals in our nervous system—regardless of how they cause various sensations and their bodily localization—form an analogous uniform phenomenon. This means that an analogy between different bodily-sensory perceptions is plausible in this "domain of cause". Such an analogy obviously relates to the bodily localization and simultaneity of all bodily sensations. Now we have to investigate whether this analogy can extend beyond the domain of bodily-sensory perceptions to the extra-bodily sensory perceptions such as the senses of sight and hearing. Just as we perceive pain or taste simultaneously localized in our body, we perceive the colour or the sound of an external object localized on the object—and in a thoroughly analogous simultaneity. However, there are no nerves in the free space that connect the brain with the perceived external object. Hence, an analogy between bodily and extra-bodily sensations relating to spatial extension, localization and temporal simultaneity (compared to the analogies between bodily sensations discussed above) seems to be a difficult or even impossible task from the outset. But the light that connects the external objects of vision with the retinal image in eyes and, thereby, with the neural network of the brain is essentially an electromagnetic phenomenon.

Would it be conceivable that the electrical network in our bodily nervous system could extend beyond the body through the analogous electromagnetism of light into immeasurable distances? The fact that the electrical phenomenon is always present in the brain and in the entire nervous system also appears to legitimize a rather speculative idea that the electrical signals could be sent beyond the limits of the physical brain into the external environment—in the form of electromagnetic short waves—and consequently extended to the extra-bodily free space and the objects located therein. Scientists have already identified brain waves, which move extremely fast as short waves and do not lose their strength and intensity even at great distances.[10] It is discovered that these electromagnetic brain waves are emitted much stronger and more intensely from open eyes.[11] Would it therefore be plausible to assume that this emission of electromagnetic waves from the brain (the extent and scope of which has not yet been adequately researched) makes our purely sensible corporeality beyond the limit of our body, more precisely beyond the limit of our brain into free space and to the objects in our environment by means of the sense of sight and hearing?

These possibilities, which have hitherto hardly been adequately researched, seem to be purely speculative, but they are necessarily presupposed and legitimized by some apparently aporetic phenomena in the extra-bodily sensibility. It is the puzzling characteristics or aporias of the sense of sight and hearing that have not yet found adequate scientific explanations; namely the visual perception of position, size and distance as well as the spatially distant objective localization of the auditory perceptions. So far, the function of light has been conceived only receptively, i.e. in relation to the mere construction of the retinal image in the eye. Now it seems to be necessary to re-examine whether light in its optical mediation between the objects seen and the physical eye has another and essential task: Does light really connect the retinal image in the eye, which is neurally processed in the brain, with the objects seen through its electromagnetism? This is obviously a question of the real unity of the eye with the visual space and, in a broader sense, of the unity between physiological optics, which begins with the entrance of light in eyes and its construction of the retinal image and is further limited to the bodily brain processes, and the purely geometrical

---

[10] See Ross, Colin A. (2010): *Hypothesis: The Electrophysiological Basis of Evil Eye Belief, in Anthropology of Consciousness*, Vol. 21, Issue 1, pp. 47–55.

[11] Ibid.

optics between the retinal image and the visual objects, which are connected to one another by the physical phenomenon of light. A possible unity between the internal and external processes in the sense of hearing can also be searched for. These and similar questions ultimately relate to a fundamental factum, namely the true essence of sensibility, which until now has been ascribed to the domain of the subject alone.

2.3 The Ontology of Senses

What is the true mode of existence of sensibility? The previously discussed epistemological function of sensibility as a mediation between the conceptually cognizing subject and the cognized object primarily refers to a connecting function, i.e. to an aesthetic nexus between the domain of the subject and that of the object. We are clear about the belonging of sensibility to the domain of the subject, but we are puzzled over its possible connection with the objects that exist independently of the subject. As mental phenomena, the modes of sensibility consist on the one hand of various qualities, such as colour, tone, taste, pain, smell, etc., which, as emphasized in the modern epistemology from Descartes to Kant, are solely subjective ideas and, as such, cannot claim objective phenomenality. On the other hand, sensibility consists of spatial and temporal structures in which the sensory perceptions extend. If we declare the spatiality and temporality—or the spatial and temporal extension—of sensibility merely to be a subjective construct, we reduce the spatiality and temporality more or less to the same status of secondary sensory qualities and deny them objective existence independent of the subject. However, space and time as qualities in the context of sensibility can hardly be equated with the purely subjective sensory qualities mentioned above, as they refer to a significantly different mode of existence. The sensation of sensory qualities necessarily takes place in spatial and temporal form and structure; i.e. the sensations necessarily presuppose their spatiality and temporality or their spatial and temporal construction—but not the other way around. Space and time seem to exist independently of sensible objectivity and are to be perceived as such. A good example would be the direct visual sensation of the free space. The spatial extension between material bodies is perceived immediately, although it has no sensory qualities such as colour, brightness, etc. The visual free space is only an empty extension, i.e. a merely *extended nothing* that is perceived directly through the sense of sight.

The purely subjective construction of spatial and temporal forms of sensibility seems from the outset to be a problematic assumption. In *Meditations* Descartes clearly differentiates sensory qualities as modes of purely mental states—as modes of thinking—from the objective existence. In doing so, however, he seems to be silent about the essential trait of sensibility, namely that of spatial extension. For Descartes ascribes spatial extension primarily to the purely physical existence—as *res extensa*—which exists independent of the soul, the *res cogitans*. Obviously, conceptual thinking or cognition has neither materiality nor extension (in other words: it makes no sense to talk about the materiality and extension of linguistic-conceptual thinking). But there is an intermediate stage between the domain of the logical subject and the domain of the materially extended body, namely the domain of senses, which is neither linguistic-conceptual nor material, but which undoubtedly has a spatial extension. This requires a revision of the Cartesian system; the extension—*extensa*—should be ascribed to sensibility in general, although it has no materiality. This would result in two different ideas of spatiality or spatial modes of existence, namely the material extension of the real object and the immaterial extension of the purely mental sensibility.

Strictly speaking, such a dichotomy of the mode of existence of space relates solely to the reality of objects and subjective sensibility and not to space itself, which can be characterized as pure extension. If, according to Descartes, *res extensa* is a real or subject-independent mode of existence of the body, then the extension of sensibility—as Descartes would have imagined it, and which his doctrine of the double mode of existence of the object (as real and mental object) would suggest—refers to a purely subjective mode of the existence of space. Is the extension of the merely mental sensibility, compared to the reality of the material extension of the body, a virtuality? We are aware of the many modes of virtuality in our mental existence, such as dreams, imaginations, reflections, echoes or even phantom pain. Does our immediate sensibility categorically belong to these virtual phenomena? How does the reality of the immediate sensibility differ from the sensible virtuality? All modes of sensible virtuality have a single factum in common that defines the ontological status of virtuality, that is the complete absence of material objectivity. We can physically approach and touch an immediately given object, but its mirror image, which lacks materiality, can neither be physically, i.e. bodily approached nor touched. Phantom pain also differs from real pain or pain localized in the body in terms of the *real* material extension of pain in the body. In this way, sensible virtuality does not exclude the reality of sensory perception.

Here we encounter a problem in the context of the ontology of sensibility: the factum of the objective presence, which gives sensibility its reality, is primarily characterized by a *real* material extension. This means that in the modes of bodily and extra-bodily sensibility a real objective extension is intertwined with the mental extension of sensibility. This merge of the bodily extension of sensibility with the material body is experienced immediately with every bodily sensation. This is indicated by the previously discussed polemic of Princess Elisabeth of Bohemia against the Cartesian conception of *res cogitans*—the immaterial and unextended soul. As implied in Descartes' answer to the princess's question, the bodily and the extra-bodily extension of sensibility does not belong to the domain of a conceptually thinking logical subject, but to the domain of a pre-logical and purely aesthetic subject.[12] It is true—in a strict Cartesian sense—that the material extension of the body is ontologically an alien factum to the essence of bodily extended sensibility, but precisely in the analogicity or rather identity of its physical extension, sensibility seems to have an infallible aesthetic-subjective nexus with the body. For the spatiality

---

[12] Thaliath, Babu (2006), *Domänen des Subjekts*, FreiDok der Universität Freiburg, Freiburg i.Br. http://www.freidok.uni-freiburg.de/volltexte/2660/, pp. 2–4.

of the material body and of the bodily extended sensibility proves ontologically to be an irreducible and final mode of existence. As such, space remains a residue that cannot be negated, as the Cartesian method of systematic negation and separation clearly shows. In reality, the Cartesian *res extensa*—within the framework of the Cartesian system itself—is just as valid for sensibility as it is for material corporeality. It is the irreducible and final mode of existence of space that guarantees the undeniable, i.e. not to be negated nexus between the material extension of the body and the immaterial extension of (bodily) sensibility.

The essence of sensibility is therefore to be determined compositionally or in a compositional nexus between the primary quality of space and the secondary sensory qualities (if we include the Lockean differentiation between primary and secondary qualities). The ontological differentiation between the subjective modes of existence of the sensory qualities and the objective-material corporeality, as undertaken by Descartes and continued in the history of modern philosophy, seems to fail precisely with regard to the spatiality that sensibility and objects perceived through senses have in common. Between the ontologically polarized entities of immaterial sensibility and material corporeality, space or spatiality seems to appear as an irreducible general factum, as space cannot be cleared out or negated either in the material body or in the immaterial yet bodily extended sensibility.

Descartes defines spatiality or spatial extension invariably as a material-physical extension; the Cartesian notion of space is derived from a primary mode of existence of the body—as *res extensa*. According to Descartes, no empty space can exist without matter. When Locke adds the attribute of material solidity to the Cartesian *res extensa*, he actually separates space from its dependence on corporeality—as conceived by Descartes—and ascribes it a completely autonomous existence from the body. In Lockean atomism, the existence of emptiness is recognized. This ontological differentiation between space and material body found its most adequate expression in Newtonian mechanics, namely in the axiomatic notion of absolute space and its completely autonomous mode of existence. Both the philosopher Locke and the natural scientist Newton seemed to imagine a threefold division of reality, that is, the soul, the body and space, in contrast to the Cartesian dichotomy between the immaterial and unextended soul and the materially extended body. The space is sublated from its Cartesian reduction—to a mere attribute of physical existence—into an absolutely primary mode of existence.

However, the problem of space in the early modern period was confined to an objective-phenomenal level. The spatiality of the mind—of the mental states and operations—was hardly regarded as an important subject of investigation in philosophy. In some well-known discourses, such as the Molyneux problem, reference was made to the spatiality of visual and haptic perceptions, but—following Descartes—they hardly seemed to go beyond the framework of a purely mental phenomenon. Berkeley's *An Essay towards a New Theory of Vision* discussed the various characteristics of the visual perception of space, as represented in the visual perception of position, distance and size of objects. Nonetheless, according to Berkeley, they are accomplished by the soul alone. The real extension of sensibility would be a complete contradiction in the context of these early modern philosophies. Descartes undertook—as part of his scientific studies—a physiological investigation of the bodily and extra-bodily sensory perceptions. He tried to demonstrate how the nervous system, which connects the brain with the whole body that includes the bodily sense organs, accomplishes the process of sensory perception on a material or physiological level. However, Descartes did not want to credit the bodily nervous system with a real extension of sensibility[13]; Instead, he tried to demonstrate a punctual nexus between the soul and the body in the brain, where all the material data transmitted through the nervous system in the body are processed into purely mental sensory states. Descartes was trying to reduce the seat of the soul in the brain to a punctual basis without extension. Although in his answer to Princess Elisabeth's polemics the connection between the soul and the body was affirmed on a pre-logical aesthetic level, i.e. within the sensibility, the idea of a real or phenomenal-bodily extension of sensibility remained alien to him. Through his physiological investigation of sensibility, Descartes apparently tried to establish a purely punctual nexus between the soul and the body and thereby to a certain extent preserve the status of the *res cogitans* originally ascribed to the soul.

Descartes' consideration of sensibility as a mode of immaterial and unextended thinking—the *res cogitans*—and subsequently its subordination to thinking[14] is strictly ontological. In his method of systematic doubt and separation, Descartes starts from the basic assumption that the sensory qualities have an ontic status that is essentially different from the materially extended body. Although Descartes recognizes from the outset the material or physical-physiological processes that underlie subjective sensibility as causal processes and makes them the subject of his scientific investigations, he remains adamant with his basic idea of the complete emergence of the sensory qualities from the material processes and their ontic autonomy and separation from the phenomenal world. Since the complete emergence of the sensory qualia—such as colour, taste, tone, warmth, etc.—obviously belongs to the domain of the subject and not to the material-corporeal world, Descartes ascribes to the sensibility—in a thoroughly problematic analogy to the states and operations of a logical subject, namely conceptual thinking, judging or cognition and also the pictorial dream and imagination—neither the ontic status of materiality nor that of extension. In short: The Cartesian system of philosophizing is based on a fundamental strategy of completely isolating all domains of the mind from the material-phenomenal reality and contrasting them ontologically with the essential features of the physical world, namely materiality and extension. This strategy surfaces in some of his philosophical positions, such as when Descartes, after a few demonstrations, employing his method of systematic doubt, finally establishes that the soul is completely different from the materially extended body and as such can exist without a body.[15]

---

[13] Descartes, René (1996): *Die Leidenschaften der Seele* (Französisch-Deutsch), trans. & ed. Klaus Hammacher, Meiner Verlag, Hamburg, p. 57.

[14] Descartes, René (1992): *Die Prinzipien der Philosophie*, trans. Artur Buchenau, Meiner Verlag, Hamburg, p. 11 (§ 32).

[15] Descartes, René (1998): *Discourse on Method and Meditations on First Philosophy*, trans. Donald A. Cress, Hackett Publishing Company, Indianapolis, p. 19.

The ontology of sensibility relates indispensably to the spatiality or to the spatial extension of the sensory perceptions. Sensations occur in the bodily and the extra-bodily extension. We always feel the sense qualities spatially; In the case of visual and auditory sensations, the free space is also perceived by the senses—as a mere perception of the distance while listening or as a real three-dimensional extension while seeing. The ontology of sensibility is therefore based on an ontological nexus between the space and the sensory qualities. We are aware that the sensory qualities are only subjective ideas and as such cannot claim material-phenomenal objectivity. Their emergence from a physiological-physical causal basis is a complete emergence and thus proves to be ontologically autonomous from the phenomenal reality. But how can we understand the emergence of the spatiality of sensory perceptions? If the spatiality of sensory perceptions is a purely subjective construct from the outset, it does not form a reality (in which the real objects perceived and the equally perceived free space participate) but a virtuality. But sensual virtuality differs from the reality of sensory perceptions and cannot exclude it, as has already been discussed.

The true mode of cognizing the spatiality of sensory perception remained an unsolved problem in the early modern (Cartesian) tradition of the theory of perception. For the Cartesian conception of *res extensa* as well as the Lockean conception of space or the spatial extension of the object as a primary quality is based on the assumption that space is a real mode of existence—that is, a mode of existence that is completely different from the perceiving and the cognizing subject. But the extension of the body, the free space and the external objects are accessible to us solely through our sensibility and its extension. How can we then distinguish the real-objective extension from the extension of the subjective sensibility in which the objects are given? As is widely known, Descartes tried to deal with this problem, which his philosophical system itself presupposed, in his two-world theory. According to this theory, things exist as real and as mental objects.[16] However, these modes of existence are merely different modes of one and the same object.[17] The unity of the object makes it difficult to differentiate a real, objective extension from its givenness in subjective sensibility, especially when we perceive this givenness in reality—and not as a virtual phenomenon. For this reason Descartes did not seem to give a clear or concrete answer to Princess Elisabeth's question about the extension of the soul in the body. Although Descartes affirms the nexus between the soul and the body in volition and sensation, he is inclined to see this nexus solely in the brain—as has already been discussed—and not as extended or spread in the whole body.

Does sensibility extend in a real space? While Descartes gives no clear answer to this question—even though he suggests the possibility of such an extension of senses within the domain of a purely aesthetic subject by recognizing the real-bodily extension of sensibility—Kant seems to view this issue in a context. For Kant, space is a necessary form of a priori sensibility (Sinnlichkeit a priori). Strictly speaking, in the Kantian system of transcendental philosophy, particularly in the Transcendental Aesthetic, spatial and temporal extension of sensibility is hardly viewed from an

---

[16] See Perler, (2004, pp. 76–77).

[17] Ibid.

objective standpoint, but rather in a subjective-transcendental context. Kant emphasizes the a priori existence of space and time in the transcendental subject—as forms of pure intuition—and thereby denies their a posteriori givenness in sensibility. The empirical intuition, in which the objects to be sensed and perceived are given, necessarily has a spatial and temporal formality. In the Kantian Transcendental Aesthetic, these forms constitute a pure intuition, which still remains an irreducible and final residue even when the empirical intuition is deprived of all intellectual attributes and sensory qualities.[18] There is sufficient reason to assume that Kant contextually restricts the discourse on space and time to a transcendental philosophical system, when he—in the transcendental discussion of these forms of sensibility—establishes space and time as necessary a priori forms of empirical or sensory perceptions. For Kant conceives of a purely objective form of existence only as a *thing-in-itself* that eludes the transcendental-subjective faculty of both cognition and intuition. According to Kant, the a priori forms of sensory perception—space and time—do not apply to the *thing-in-itself*. That is, the *thing-in-itself* exists neither in space nor in time, which, according to Kant, are merely a priori forms of pure intuition and, as such, can only belong to a transcendental subject.

Had Kant tried to answer Princess Elisabeth's question (provided that this question necessarily relates to bodily sensibility), he would recognize the spatial extension of sensibility, but at the same time emphasize that the extension of senses does not owe its origin and existence to a *real* bodily or extra-bodily extension (a posteriori), but solely to a transcendental-aesthetic subject. That is to say, according to Kant, a material-bodily or extra-bodily extension cannot impart the sensibility its spatial extension; the phenomenal world only provides contents of sensation; it is only through a transcendental-aesthetic subject that they are a priori intuited in space and time, which are a priori existing in subject. Intuition is accordingly a process of a priori-subjective spatialization and temporalization of objects given in sensations (Empfindungen).

In doing so, the spatial and temporal modes of existence pertaining to objects and independent of the transcendental subject do not seem to be sufficiently taken into account. However, Kant has the fundamental notion that objects are given in empirical intuition. Although Kant considers this givenness as an appearance again in a transcendental-philosophical framework, it seems to presuppose a residual factum of a purely objective givenness, to which the Kant's theory of affection as the primary instance of cognitive process (although Kant identifies the source of this affection in "thing in itself") and his programmatic philosophy of synthesis—as a transcendental doctrine of binding knowledge, and also of sensory perception—refers. The givenness of the object inevitably posits a givenness in the objective form of space and time—i.e. in the purely objective spatiality and temporality. Kant would fundamentally deny such a conclusion and consider a purely objective existence independent of the subject to be a *thing-in-itself* that—in contrast to appearance (Erscheinung)—does not *exist* in space and time.

[18] Kant (1990, p. 64).

However, if we can philosophically accept a world of material objects that exist independently of the cognizing and perceiving subject—just as we legitimize the existence of our body—we would recognize the essential features of this objective existence as spatial and temporal extension. Accordingly, in empirical intuition, we ascribe to the givenness of objects the essential traits of spatiality and temporality, which, as primary qualities, are independent of the subject. If—following Kant—the empirical intuition forms a transcendental-aesthetic-subjective and as such a priori spatialization and temporalization of the objects given in sensibility, it can be eventually viewed as a (subjective-sensory) spatialization of spatiality and temporalization the temporality of the *given* objects. However, such an idea goes beyond the framework of a (Kantian) transcendental aesthetic.

In the introduction to the *Transcendental Aesthetic*, Kant mainly discusses the a priori forms of sensibility, namely space and time. Further considerations about space refer primarily to the geometry or to the apriority of axiomatic geometrical intuitions. The geometric shapes are objects of sensible, i.e. visual perception, but the space of geometry differs from the immediate spatiality of sensory perceptions—especially of sight—in many ways. Strictly speaking, the geometric space forms a theoretical space, whereas the direct sensory perception extends in an aesthetic space. However, the free space, which is merely an extended nothing, remains unchanged in the geometrical intuitions and in the immediate sensibility. In the case of geometric forms that are imagined in free space, it is hardly possible to distinguish between their intuitive conception and the immediate sensory perception. An elementary geometric form such as a line or a surface appears in its true or axiomatic form both in its intuitive imagination and in the sensory perception (particularly in visual perception). But if we intuitively visualize the axiom that "space has three dimensions" by means of three coordinates—namely the abscissa, the ordinate and the elevation—or directly see it in a material, i.e. in an architecturally objectified form, we will not be able to visually perceive the exact axiomatic form or structure of space coordinates. For our direct vision and our imagination, which can only occur visually, are perspectively preconditioned.[19] This means that our immediate visual space and our—equally visual—imaginary space are based on a perspectival spatial structure. The perspectival structure constitutes the most important characteristic of the visual space, to which we have access only through sight. Due to the perspectivity, the true structure of the spatial coordinates can neither be directly *seen* nor a priori imagined *in their entirety or perfection*—as three axes intersecting at one point forming right angles. For the structure of the spatial coordinates appears to us perspectively deformed, both during direct vision and in our intuitive imagination or visualization. We can only see a few angles in the shape of the right angle. The other angles around the origo (zero point) appear to us as acute and obtuse angles. The fact that all angles that construct the spatial coordinates at the origo are right angles corresponds to the reality of the spatial coordinate structure and the axiomaticity of Euclidean space mentioned above, but we must derive this knowledge—as a theoretical-geometrical

[19] See Thaliath, Babu (2010): *Natur und Struktur der Kräfte*, Verlag Königshausen & Neumann, Würzburg, p. 48.

construct—from a basically perspectival appearance. This would suggest that the intuitive-axiomatic conception of this axiom requires a theoretical-spatial construction—as a necessary correction—in addition to the immediate visibility of the spatial coordinates in an aesthetic-perspective visual space.[20]

Kant's preference for geometrical-theoretical space within the framework of a transcendental aesthetic seems to be of a strategic nature—in order to justify the a priori nature of space programmatically in a transcendental philosophical system. For Kant uses the context of a transcendental aesthetic primarily to demonstrate his basic conception (on which he builds his transcendental philosophical system), that is, the a priori status of synthetic judgements, as represented above all in mathematics and in the mathematical sciences such as mechanics. "How are synthetic judgments a priori possible?" is the basic question posed and discussed in the introduction to the *Critique of Pure Reason*. The apriority of spatial conceptions (Raumvorstellungen) appears clearly in the geometrical-axiomatic intuitions. The fact that space and its geometric properties, as axiomatized in Euclidean geometry, are treated within the framework of a *Transcendental Aesthetic* seems to be justified by the fact that the geometric forms and structures are primarily objects of aesthetic or visuality. However, precisely during seeing—both in the immediate visual perception as well as in the intuitive-visual imagination—a perspectival visual structure emerges in the first place, which is geometrically constructed, but which, in comparison to the geometrical intuitions, appears to be objective or phenomenal. For the perspectival structure of vision and visual space is determined by the real light rays which are reflected by the objects and converging on the eye point. If we discuss the spatiality of seeing—as the most important mode of empirical intuition—in the context of a *Transcendental Aesthetic*, we must start from a perspectival spatial structure that underlies both the immediately perceptible visual space as well as our a priori visual imagination.[21]

The mode of existence of visual perception consists of a composition of the perspectival space structure, which extends into the free space and to the objects, as well as of the sensory qualities such as colour, brightness, etc., in which the objects appear. As previously discussed, these sense qualities are purely subjective ideas,

[20] See Thaliath, Babu (2005): *Perspektivierung als Modalität der Symbolisierung. Erwin Panofskys Unternehmung zur Ausweitung und Präzisierung des Symbolisierungsprozesses in der Philosophie der Symbolischen Formen von Ernst Cassirer* (Dissertation), Verlag Königshausen & Neumann, Würzburg, pp. 287–288.

[21] The complete absence of a discussion of the perspectival structure of visual space in *Transcendental Aesthetic* could be a coincidence, but the discourse on the origin of perspectivity in vision has a long pre-Kantian history. Ever since the geometric construction of the central perspective was discovered by Filippo Brunelleschi and was further developed as a technique of painting by Leonardo da Vinci, Leon Battista Alberti and others, various theories of perspective emerged in the Renaissance and Early Modern Period, which in principle related to the real structure of vision. In addition, the emergence of visual space perception became the subject of various philosophical endeavors—by Locke and Molyneux, as represented in the famous *Molyneux's Problem*, by Berkeley in his major work *An Essay towards a New Theory of Vision*, and by the French philosophers Diderot, Condillac and Voltaire. The historicity of this discourse seemed to be largely suppressed by Kant in the context of his *Transcendental Aesthetic*—especially in his statement "Space is a necessary a priori notion" ("Raum ist eine notwendige Vorstellung a priori". See Kant (1990, p. 67).

although they are objectively, i.e. physically and physiologically caused. However, they are extended objectively in the direct vision. Can we ascribe a purely subjective mode of existence—that is, a mode of existence in which the objects seen do not directly participate—to the perspectival spatial structure of seeing in analogy to the qualities of the sense of sight such as colour or brightness? The complete ontological separation of vision from the objective domain of the body and the extra-bodily objects would mean that the entire visual perception—i.e. the secondary qualities of the objects seen and the perspectival extension of the visual space—is a subjective construct. What connects the eye with the objects seen and with the free space (not directly *seen*, but visually perceived) are the light rays that are reflected from the objects and converge on the eye point and, thereby, create the images of objects on the retina. If the subject has only the two almost identical retinal images in both eyes as the sole input during the visual process, it should develop the entire visual perception *virtually* through these retinal images alone. However, this would mean that the subject develops an immeasurably large perspectival visual space from a comparatively very small retinal image, in which the visual objects are reversed and—according to the shape of the retina—spherically distorted. As we experience in our direct vision, the closer objects appear in approximately correct size and distance, but also the distant objects such as a mountain or the sky, which *exist* in huge extension and distance, although their images on the retina are relatively too small. In addition, the spatial distances of the objects seen and their perspective-dimensional depth are not represented on the two-dimensional retina—a fact that Berkeley emphasizes in his treatise *An Essay towards a New Theory of Vision* in order to demonstrate the inconsistency of the geometrical-optical explanation of the visual distance perception. That the perceiving subject develops a field of vision with its immeasurable size and depth *virtually* from a tiny retinal image, i.e. independent of the *real* objects seen and the equally real and visually perceived free space, seems to elude our logical understanding. But if, conversely, we assume that the visual perception of size, distance and position are based on the *real* size, distance and position of the objects viewed and the real extension of the visual free space, we tacitly assume that our vision is *really* and immeasurably extended in a huge perspectival visual space that embodies both the objects and the free space *seen*. Obviously, such an idea would elude our understanding. So we get into the difficult position of having to choose between these two possibilities, i.e. to accept one of them as truth.

The incredibility of the idea that our sensibility, i.e. our visual and auditory sense can *really* extend extra-bodily, also lies in the fact that we do not perceive the presence of a material medium between us and the objects seen and heard. We are *aware* of the bodily extension of all tactile sensations, such as cold, warmth, wetness, etc., as well as pain or taste to a greater extent. In contrast to it, the visual and auditory sensations extend into an extra-bodily space and are localized mostly on objects outside of the body. The external, objective localization of the visual and auditory sensations as well as the visual perception of the external free space clearly form *aporias* of our sensibility. We tend to look for a solution to these aporias within the framework of a philosophical subject-transcendentalism (as in Kant) or within the framework of natural sciences such as neurobiology. The reduction of the sensibility and its spatial extension to a transcendental subject or to the neural processes in the brain, through which we try to *solve* these aporias, excludes the possibility that our sensibility can extend in a *real*—bodily and extra-bodily—space.

References

Descartes, René (1972). *Meditationen.* Artur Buchenau (Trans. & Ed.). Meiner Verlag. Hamburg.

Descartes, René (1992). *Die Prinzipien der Philosophie*, Artur Buchenau (Trans.). Meiner Verlag. Hamburg.

Descartes, René (1996). *Die Leidenschaften der Seele (Französisch-Deutsch).* Klaus Hammacher (Trans. & Ed.). Meiner Verlag. Hamburg.

Descartes, René (1998). *Discourse on Method and Meditations on First Philosophy.* Donald A. Cress (Trans.). Hackett Publishing Company. Indianapolis.

Descartes, René (2003). *Meditations and Other Metaphysical Writings.* Desmond M. Clarke (Trans.). Penguin Classics. London.

Descartes, René (2009). *Meditationen.* Christian Wohlers (Trans. & Ed.). Meiner Verlag. Hamburg.

Ebbersmeyer, Sabrina (2015, Ed.). *Der Briefwechsel zwischen Elisabeth von der Pfalz und René Descartes.* Wilhelm Fink Verlag. Paderborn.

Kant, Immanuel (1990). *Kritik der reinen Vernunft.* Raymund Schmidt (Ed.). Meiner Verlag. Hamburg.

Kant, Immanuel (1998). *Critique of Pure Reason.* trans. & ed. Paul Guyer and Allen W. Wood. Cambridge University Press. Cambridge.

Lauth, Bernard (2006). *Descartes im Rückspiegel.* Paderborn.

Maier, Anneliese (1968). *Zwei Untersuchungen zur nachscholastischen Philosophie.* Rom.

Perler, Dominik (2004). *Inside and Outside the Mind—Cartesian Representations Reconsidered.* In: *Perception and Reality. From Descartes to the Present.* Ralph Schumacher (Ed.). Mentis Verlag. Paderborn.

Ross, Colin A. (2010). *Hypothesis: The Electrophysiological Basis of Evil Eye Belief*. In: *Anthropology of Consciousness.* Vol. 21. Issue 1.

Thaliath, Babu (2005). *Perspektivierung als Modalität der Symbolisierung. Erwin Panofskys Unternehmung zur Ausweitung und Präzisierung des Symbolisierungsprozesses in der Philosophie der Symbolischen Formen von Ernst Cassirer* (Dissertation). Verlag Königshausen & Neumann. Würzburg.

Thaliath, Babu (2006). *Domänen des Subjekts.* FreiDok (Freiburger Dokumentenserver of the University of Freiburg). Freiburg i. Br.

Thaliath, Babu (2010). *Natur und Struktur der Kräfte.* Verlag Königshausen & Neumann. Würzburg.

Chapter 3
The Extension of Senses

3.1 The Problem of Emergence in Sensibility

The aporia of sensibility can mainly be traced back to its emergence from the material phenomena. Sensibility is an entirely mental phenomenon emerging from a phenomenal causality; its emergence from the extra-bodily and the bodily-phenomenal causal chains is not a partial but a complete emergence. This complete emergence is at the same time a complete ontological leap—from the domain of the material phenomena to the domain of the pre-verbal mental states. The material-phenomenal causation, the effects of which constitute the entire modes of sensibility, extends both bodily and extra-bodily. While the nervous system, which extends throughout the body, enables the bodily localization of sensations such as pain, warmness, taste, etc., the visual and auditory sensations extend outside the body; their material mediality, which is to be analogized to the material body, appears to us as a riddle. In this way, in our sensibility, a purely material-phenomenal causal domain can be ontologically differentiated from an immaterial-mental domain of effect or sensible reality. This causal material phenomenality underlying sensibility refers physically to the entire nervous system that is spread throughout the body, and is built on neuronal states and processes; it seems to extend externally or extra-bodily to various natural phenomena such as light and air waves, chemical properties of the seen surfaces, mechanical vibration of bodies (which generate the air waves), etc.

While the material phenomena give rise to a completely ontological emergence in their generation of purely mental sensibility, this causation forms fundamentally an ontological causality. The principle of such causality is that a higher and more complex phenomenon is caused through the elementary phenomena. In the normal case of ontological causality, the real phenomenon or the phenomenon individuated as reality includes within itself its elementary ontological causality; such as the molecules, whose reality emerges from an elementary-ontological atomic causality, or the reality of the atoms, which have furthermore an inherent subatomic

B. Thaliath, *The Embodiment of Senses*, https://doi.org/10.1007/978-981-96-8468-7_3

causality.[1] The emergence of mental sensation, however, seems to have an existence and phenomenality totally divorced from all causal material phenomena, especially from its neural basis. The colour of an object is ontologically a completely different mental mode of existence compared to all material-causal phenomena. Likewise, the taste or sound of an object is distinct from all the physical phenomena that cause those mental states. However, this ontological separateness and autonomy of a mental sensation is limited to its sensory content, represented in all sensations of colour, sound, noise, taste, pain, etc. As has already been discussed, these modes of sensibility do not arise as mere sensory content, but also in certain forms, the formality of which necessarily comes about through a spatial structure. The causative material phenomena, from which the mental sensory content arises in a completely ontological separateness, seem to participate precisely in the emergence of the spatial form and structure of sensibility. This means that the bodily and extra-bodily extension of senses does not only seem to come about mentally, but rather phenomenally or through the physical phenomena that cause it. Consequently, the nexus between the extension of the various modes of sensibility and the extension of their physical-causal phenomena needs to be determined more closely.

With regard to spatial extension, the sense of sight proves to be the most enigmatic of all modes of sensibility. For in the visual perception of space, a very clear difference develops between the reality or the real form and structure of vision and its neuronal causation. The structure of visual space that we experience seems to be constructed directly by the objective phenomena, such as the material objects and the light rays converging on the eye point, as well as by the free space. Especially the closer objects appear to us in approximately correct size, solidity and spatial distance—that is, more or less in the same way they are or exist in space. Likewise, the perspectivity or perspectival structure in the visuality of the objects and in the perceptibility of the open free space arises from a real phenomenal structure, namely the convergence of the rays of light reflecting from all objects onto the eye point. However, following Kantian transcendentalism, we would state that this visual perception of space is fundamentally a subjective a priori idea, or—to use the words of neurobiologists—that it arises solely from the neuronal processes in the brain. Both explanations are causal explanations; the former from philosophical epistemology and the latter from neurobiology. Here are some essential problems to identify in visual spatial perception: What is the very first basis or original causality of visual space perception, i.e., visual perception of size, distance, position and free space? Does visual space perception originally occur on the *level or domain of effect*, i.e., in the *real* visual space itself, whereby the neuronal processes do not bring about the visual space perception as the sole originator, but rather as a causal correlate or participant—like a "support system"? This question clearly builds on a scepticism as to whether the true function of neuronal processes (which in the normal case are thought to be purely causal) is a *complete causation* or the mere maintenance of a visual-spatial

[1] On *ontological causality*, see Thaliath, Babu (2009): *The ontological causation*, Journal of Dharma 33 (2008), Bangalore, pp. 33–56.

perception that has already arisen on the level of effect or in real physical space.[2] If the neuronal processes in direct vision do not have a solely causal but rather a correlative- or participatory-causal function, they form a certain operational unity with the real visual space or with the geometrical-optical structure of vision.

The aporia of visual perception refers not only to the visual perception of position, size and distance, but also to the emergence of visual images from the neuronal processes which, as material phenomena, have a completely different ontological status compared to the purely mental reality of seeing, as previously discussed. The complete ontological emergence of the sense of sight—of sensibility in general—from material phenomena makes it difficult to provide a causal explanation of their existence or their existence in the sensory content, such as colour, and in the spatial-temporal form and structure. The problem of an adequate causal explanation of sight can be introduced by asking a few questions:

1. Does the sense of sight arise merely neuronally, where the brain has only the retinal image (or the retinal images in both eyes) as input?
2. Can the sense of sight, when it arises through the neural processes in the brain, actually extend to a real visual space that consists of real objects and free space and that is perspectively structured by the light rays converging on the eye point?

Modern philosophy as well as neurobiology and other related sciences answer the first question in the affirmative and the second in the negative. According to them, the emergence of the entire visual experience is attributed solely to the neuronal processing of the two almost identical retinal images (in both eyes). The visual image and its content-related, formal and structural characteristics and qualia—such as the colourfulness or brightness as well as the perception of the size, position, free-space distance and solidity of the objects seen, the perspectivity, etc.—arise solely from the neuronal processes. The possibility of the unfolding of the sense of sight, as indicated in the second question, although it solves all the aporias of visual spatial perception mentioned above, seems to defy ordinary common sense. For it implies the *real* expansion of the sense of sight through immeasurable distance, width and height. That our sense of sight really extends out of the eye and thereby incorporates our extraordinarily far-reaching visual image—which reaches to the distant horizon, to the height of the sky, etc.—seems incredible to us from the outset. However, it is primarily not the lack of evidence, but the implausibility that prevents us from assuming that our sense of sight can achieve such a thing.

However, the first possibility accepted by modern philosophy and neurobiology, that our sense of sight and our sensibility in general arise solely through the

[2] It cannot be denied here that the neuronal processes form the original and necessary causal basis for visual spatial perception, but due to their confinement in the brain or its materiality, they seem to be rather an accessory or accompanying causation that—as an external causal basis—correlates on *extra-bodily level or domain of effect* with direct vision and its geometric-optical principles. Such a causal correlation necessarily occurs in a unity of the visual space with the eye on a perceptual-operational level, which presupposes the unity or the synthetic nexus or connectedness of the physiological optics with the extra-bodily geometrical optics.

neuronal processes in the brain, seems equally questionable to us. For here a material phenomenon causes purely mental states and processes; the entirely ontological difference between the domain of cause and the domain of effect prevents from the outset an ontological-causal reduction of mental sensibility to its neuronal causality. For each of the visual traits discussed above and for the other modes of sensibility, a specific neuronal origin in the brain could be identified. Our direct experiences of sensory perception prove that these material causes only allow mental states to arise. Ultimately, however, we must "believe" that they are sufficient causes of our mental states of sensibility. For these material causes have ontologically nothing in common with their mental effects, which constitute the whole of the above-mentioned traits of sensibility. The massive appearance of a mountain is supported by a certain neural state in the brain, but it is far from conclusive to state that the gigantic appearance comes about through this neuronal causation alone. For the reality of this immediate visual experience of ours seems to occur first and foremost on the level or domain of effect itself, especially if we take into account its immeasurable spatial extent. Likewise, it can hardly be explained exclusively neuronally that we visually perceive the true objects in our near field of vision in their approximately correct and completely different sizes and distances—a problem that many philosophers of modern times such as Berkeley, Condillac, Helmholtz and others have dealt with. The differences in the size of the retinal images do not provide a sufficient reason for the completely different visual size perceptions of the real objects; the distance of the objects seen and their differences are not represented on the retina (a fact that Berkeley emphasizes in his famous and still topical treatise "An Essay towards a New Theory on Vision").

We wonder how the relatively small and purely material states in the brain can give rise to an essentially different mental reality of our immense visual field and the other modes of sensibility. But the brain produces the mental virtuality, such as memory, dream and imagination, in which the real objects do not directly participate. The inevitable neuronal causation of this sensible virtuality also proves the potential of the brain to develop purely mental states solely from the material processes. However, this does not overcome or invalidate the aporia of immediate sensibility. For the spatial and temporal forms, dimensions and structures of the immediate sensory perceptions do not seem to be developed just from the brain, but primarily obtained from the real phenomena or the objects of the sensory perception that are actually present. The aporia of sensibility seems more comprehensible to us when we ask a basic question about the true *reference* of the sensible perception of extension, distance, solidity of the (perceived) material objects as well as the extension and structure of the free space perceived through the sense of sight or hearing. *How does* the brain *know* about the roughly correct size, distance, solidity and position of objects (especially nearby objects) in the visual field without the brain having a *direct reference* to the real objects seen and to the visually perceived empty space?[3] Such a necessary reference—and thus direct *access* to the real objects seen—would

[3] Thaliath (2005, p. 232).

relate to the second question posed above, which, however, hardly saves us from our aporetic understanding of sensibility.

As mentioned earlier, the possibility of sensory virtuality teaches us that sensory qualia such as colour, sound, taste or tactile sensations are developed by the brain (according to neurobiology) or by the subject (according to philosophical epistemology). All these qualia form what Locke called secondary qualities. Although the neuronal causation and emergence of these sensory qualia seems puzzling to us (due to the entirely ontological leap from material to mental phenomena), it is the primary qualities of sensibility, especially the spatial and also the temporal dimensions, that pose the greatest riddles to us. These primary qualities of sensibility contain all the essential features of spatial and temporal extension, such as size, distance, position, simultaneity, movement, etc.[4] The mere existence of these spatial and temporal dimensions of extra-bodily sensibility in the mode of a retinal image in the process of seeing and the vibration of the eardrum in the process of hearing is quite inadequate for the subject to develop the proper extension of sensibility. If the process of seeing and hearing begins only with the bodily input, that is, with the retinal image and the vibration of the eardrums—in other words: if in the neuronal development of seeing and hearing all extra-bodily factums, such as the rays of light converging on the eye point, the free space, the real objects in the visual field, the air waves, etc., are excluded—, it seems to us inexplicable from where the brain gets or gains

[4] The sensory—primarily visual—perception of movement is the perception of a spatiotemporal extension of a real object. The movements of objects in visual field are depicted—following the laws of geometrical optics—only inappropriately on the retina of the eye:

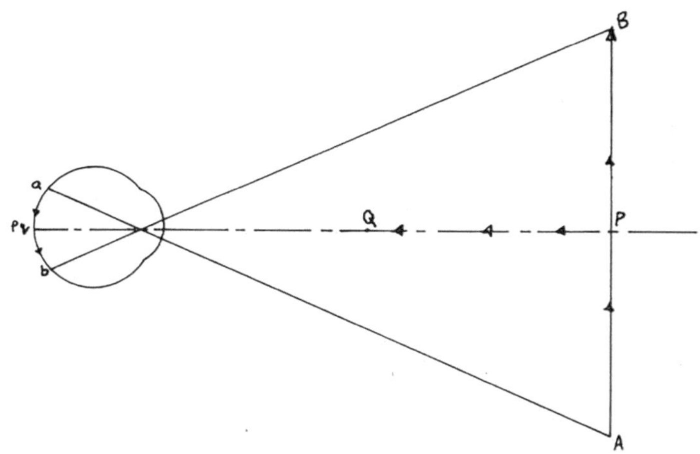

Fig. 3.1 The optical perception of movement

The movement of an object horizontal to the retinal plane (from A to B in Fig. 3.1) is imaged on the retina in the opposite direction and very minutely (ab). When the object moves in the direction of the eye point—along the visual axis (PQ)—its movement is not spatially imaged on the retina. The tiny magnification of the image here is not an adequate basis that could lead us to the correct perception of the spatiotemporal extension of this movement.

all *references* to the approximately correct spatial and temporal dimensions of these modes of extra-bodily sensibility. Obviously, in this way, the visual and auditory perception of the correct spatial and temporal dimensions of the environment form an aporia. The aporia of the extra-bodily sensibility is primarily a *phenomenal aporia* that relates to the limits of neural processing of bodily inputs in vision and in hearing.

Aporias arise subjectively, or more precisely, subjective-epistemologically; they mark the limits of the subjective understanding. But they owe their origin to objective phenomena. For it is fundamentally the phenomenon that sets limits to the subjective understanding or thinking. Aporias arise when the object of understanding appears enigmatic to the subject.[5] The origin and boundary of space as well as the beginning and end of time form aporias, where the *aporetic nature* of their form and structure is ultimately based on the nature and mode of existence of these objects of understanding themselves. In short: the subjective aporias *originally* arise through the objects; it is the aporia of *phenomenal individuation* that sets limits to the subjective understanding or cognitive faculty and *arrests* it in that aporia. However, the aporias also arise from the subjective reluctance that the subject cannot simply accept the insights *dictated* by the phenomena as they seem *implausible*. The fact that time can have no beginning and no end is a realization that the nature and mode of existence of time itself dictate to the subject. In the same way, space seems to dictate its boundlessness to the subject. But the subject is hardly prepared to merely accept such realizations, as they elude the subjective power of imagination. Since the subject cannot *imagine* such realizations, it attempts to reinterpret them as aporias or antinomies. However, in addition to the impossibility of imagination and implausibility, there are also other sources of aporia; e.g. subjective pride, which can develop into egoism (since the subject does not *want* to merely accept the knowledge dictated by the phenomenon, which is an enigma or aporia) or certain strategies for marginalizing or paradigmatically suppressing the objectively given premise, etc.

When fixing the aporias, a certain subjective stance to epistemologically reject or overlook the existence of phenomenal aporias is often identified. In contrast, the subjective acceptance of phenomenal aporias could lead to fruitful insights and possibly to the resolution of aporias. Aporias of sensibility also arise when the investigation or search for causal explanations is limited to a particular field of science and *all the essential characteristics* of sensible reality—such as the spatial-perspectival

---

[5] See Thaliath, Babu (2016): *Wissenschaft und Kontext in der frühen Neuzeit*, Verlag Karl Alber, Freiburg i. Br., p. 280: 'The original ideas of aporia in the philosophy of Plato and Aristotle point to that borderline experience in thinking, that is, to the hopelessness in the thought process in which one is primarily seeking a solution. However, Aristotle points out that the hopelessness in thinking can be traced back to the aporetic nature of the matter or the object of thought: „Wer einen guten Weg finden will, für den ist es förderlich, die Ausweglosigkeit gründlich durchgehalten zu haben. Denn der spätere Weg ist die Lösung dessen, worin man zuvor keinen Weg hatte. Man kann nicht lösen, wenn man den Knoten nicht kennt. Wenn man aber im Denken keinen Weg hat, dann zeigt das diesen Knoten in der Sache an.“ (Aristoteles, Met. B 1, 995 a24–b4) See Jacobi, Klaus (2008): *Kann die Erste Philosophie wissenschaftlich betrieben werden? Untersuchungen zum Aporienbuch der aristotelischen» Metaphysik«*, in *Metaphysisches Fragen. Colloquium über die Grundform des Philosophierens*, hrsg. von Paulus Engelhardt und Claudius Strube, Böhlau Verlag, Collegium Hermeneuticum, Bd. 12, Köln – Weimar – Wien, p. 31 ff.

extension of the visual field or the extra-bodily-objective localization of auditory perception—are discussed in this field alone. Extending the *causal basis* to the reality of senses would then imply the extension of the process of sensibility to other fields of science, such as geometrical optics. Compared to neurons, this has a different phenomenal basis, namely the phenomenon of light. Such an extension presupposes that the actual visual process takes place in a unity of bodily-physiological optics with extra-bodily geometrical optics. This unity clearly implies a unity of the physical eye with the extra-bodily visual space. In the following, we will examine how this perceptual unity of extra-bodily sensibility is determined by the *essential characteristics* of the actual process of seeing and hearing.

3.2 The Aporias of Visual Perception

3.2.1 The Retinal Image

The retinal image in the eye is the boundary between two domains of the optics of vision, namely between purely physiological optics and extra-bodily geometrical optics. If our direct visual image is created solely by neuronal processes in the brain, the retinal image—or the retinal images in both eyes—is the only input in the course of vision that is provided by the extra-bodily geometrical optics. However, the optic nerves do not transmit this input to the brain as images, but in the form of electric signals. This means that the very first transmission of the retinal image in the neuronal processes involved in the course of vision in the brain is subject to a radical *phenomenal* transformation. The fact that the light rays converging on the retina act on the light-sensitive surface of the retina and thereby generate the photoelectric signals supposedly marks the final stage of the geometric-optical process, which alone connects the eye with the real objects and with the free space in the field of vision.

However, considering the retinal image as an image—or as an analogue of photography—seems problematic from the outset. This is because the retinal image is *not seen* in the process of vision. As long as the retinal image is not directly seen with the naked eye in the process of vision, it cannot be said to have many of the characteristics of an ordinary image, such as colour and perspective. That is to say, only the factum of *subjective* vision is able to transform the image on the retina into the status of a real—coloured and perspectival—image. Without subjective vision, the image on the retina remains just an exposed layer on which objects appear colourless and without a perspectival structure, with zones of light and shadow overlapping and the visual free space not being depicted. The *real* or phenomenal image on the retina is a light phenomenon and not an image phenomenon that presupposes subjective vision.

The myth of the retinal image was emphasized by philosophers and scientists of optics in the developmental history of theories of vision and physiological optics. In

"Dioptrics", Descartes disputes the idea that the retinal image is visually sensed in the eye:

> Wenn nun auch dieses Bild, das auf diese Weise ins Innere unseres Kopfes gelangt, immer noch eine Ähnlichkeit mit den Gegenständen behält, von denen es ausgeht, so darf man sich doch, wie ich es Ihnen ja schon verständlich genug gemacht habe, die Sache nicht so vorstellen, als ob wir durch diese Ähnlichkeit eine Empfindung von ihnen bekommen, als ob es noch andere Augen in unserem Gehirn gäbe, durch die wir sie wahrnehmen könnten.[6]

In his work "Depth Perception Through Motion", Myron L. Braunstein cites a similar observation by Kepler:

> Kepler (1604) leaves to the natural philosopher the question of whether the retinal image "is made to appear before the soul or tribunal of the faculty of vision by a spirit within the cerebral cavities, or the faculty of vision, like a magistrate sent by the soul, goes out from the council chamber of the brain to meet this image in the optic nerves and retina, as if it were descending to a lower court.[7]

Both Kepler and Descartes emphasize the fact that the retinal image, which is not seen, has a significantly different status as input in the process of vision. Here, we can only speak of a similarity between our actual vision or visual image and the retinal image in a restricted sense. This is because the objects are not depicted in their correct position and size on the flat, i.e., two-dimensional retina. That is, in the geometric-optical process of creating the retinal image, the static phenomena and the movements are reversed and depicted in an extremely diminutive way. However, the visual depth perception, to which the retinal image provides no input data, seems to be here the most difficult problem or aporia:

> The problem of visual perception has a long history. For hundreds of years men have felt the need for some explanation of why things are seen. Among the many puzzles to which the problem leads, perhaps the oldest and most general of all is this: how can one account for the richness of sight considering the poverty of the image within the eye? Vision depends on this retinal picture. But what an inadequate thing the image seems to be when compared with the result! The visible scene has depth, distance, and *solidity;* the image is flat. How can vision depend on the pictures in the eyes and yet produce a scene which extends to the horizon? The physical environment has three dimensions; *it* is projected by light on a sensitive surface of two dimensions; it *is* perceived nevertheless in three dimensions. How can the lost third dimension be restored in perception?[8]

Philosophers and psychologists, as well as other scientists in the field of ophthalmology, may point to various facts, such as the haptic perception of size, the movement of the pupil and of objects in the visual field, perspectivity, stereoscopic or binocular vision, etc., from which the correct visual perception of position, size and depth could arise. But all these "cues" can be methodically negated without the correct spatial vision ceasing to exist. The factum of stereoscopic vision contributes to the perception of depth, but it only perfects this already existing characteristic of vision.

---

[6] Leisegang, Gertrud (1954): *Descartes Dioptrik*, Meisenheim am Glan, p. 99.

[7] Braunstein, Myron L. (1976): *Depth Perception Through Motion*, New York 1976, p. 5.

[8] Gibson, James J. (1904): *The Perception of the Visual World*, Greenwood Press, Connecticut, p. 2.

The origin of visual depth perception—like the origin of the correct perception of position and size in vision—is to be found elsewhere.

The immediate logical conclusion that can be drawn from the problem of the invisibility of the retinal image in the process of vision would be as follows: Since the retinal image, constructed through a geometric-optical procedure, is not seen in the process of vision, it primarily forms a mere nexus between the physiological part and the (extra-bodily) geometric-optical part of the eye optics. This means that the retinal image has a connecting function in addition to a receptive function.[9] The unity of the physical eye with the extra-bodily visual space and thus the unity of physiological with geometrical optics in the process of vision, which seems to be based on the retinal image and its connecting function, will now be examined in detail.

Seen with the naked eye, the image on the retina resembles the image in a camera—except for the spheroidal form or concavity of the image on the retina. But if the factum of direct vision is missing, some essential features such as colour and perspective disappear, as discussed earlier. However, some light-related properties, such as sharpness and contrast, remain. If the image on the retina is blurred—as in the case of short-sightedness or long-sightedness—our visual image also appears blurred to us. Similarly, the cylindrical curvature of the retina (astigmatism) has a direct effect on our visual image. All these optical anomalies of the eye can be corrected geometrical-optically, or by using appropriate glasses. This means that the essential characteristics such as line type, contrast, sharpness, etc. (which apparently are the essential characteristics of the black and white image) remain in the retinal image, even if we cannot see it. Because these characteristics are *created* solely by the light on the light-sensitive surface of the retina—by its sharpness, its intensity in brightness or its absence in shadows, as well as all the *gradations* in between. We will now examine how the position and size of objects imaged on the retina relate to the actual objects or the approximately correct position and size of the objects in our immediate field of vision.

3.2.2 Visual Size Perception

Objects and their movements appear very small on the retina. Both large and close objects, such as a high-rise building, trees or rivers, and distant objects, such as the sea and sky stretching out to the horizon, appear very small on the retina. Likewise, the movements or the spatiotemporal extension and speed of the movements appear extremely reduced in proportion on the retina. Our immeasurably extended visual space is said to be developed from its extremely small image on the retina (which has the size of the inner surface of a hemisphere with a diameter of 2.5 cm). This presupposes, first and foremost, that the brain, which processes the input of the retinal image neuronally and presumably develops the real sense of sight from it, is

[9] Thaliath (2005, p. 209ff.).

supposed to have the appropriate references—and not only to the correct sizes of the real appearances in the field of vision, but also to the immeasurable extension of the free space in the field of vision. The very small image on the retina offers at most a certain proportional representation of the objects—according to the principles of geometric optics—but it offers no references to the real, directly seen sizes in our field of vision. The extension of the open free space reaching to the horizon and the sky is not depicted on the two-dimensional plane of the retina either; there the reference is completely absent.

In the history of the development of the optics of the eye, philosophers and scientists in the fields of optics and ophthalmology have proposed various "cues" that could serve as references for correct size perception in the process of vision. It was assumed that familiarizing oneself with the bodily-tactile size perception of objects during the process of vision could suggest the correct (visual) size perception. However, this assumption is problematic from the outset, as our bodily-tactile perception of size is limited to the few close and relatively small objects; large phenomena such as a high-rise or a tree, as well as the vast expanse of a mountain or sea, cannot be perceived haptically.[10] It is equally inconclusive to claim that a tactile *sense* of size is merely transferred to the visual perception of size. It has been claimed that during the process of vision, certain a priori "ideas" of the sizes of objects could serve as references for an approximately correct visual perception of size. Such a philosophical-epistemological idea would prove problematic because there is no single idea of the sizes of objects. The representational genres, such as plants, stones, trees, rivers, buildings, etc., vary in size—just like the idea of a human being, that includes both short and tall men and women. Furthermore, it is completely absurd to assume that an abstract idea or concept of size is *transformed* into *actual* visual sizes during the process of vision.

While searching for the true references for the correct, i.e., immediate visual perception of the object's size, we will certainly realize that visual perception of size is one of the most appropriate proofs of the aporia of visual perception. In the case of such an aporia, the solution seems to be dictated to the subject by the objects themselves: The true and only reference for an approximately correct visual perception of the size of the object is the *real object itself,* that a person encounters visually. However, we hesitate to simply accept such a dictation of knowledge from the object, even though it would solve all the aporias regarding the visual perception of size—of static and moving objects. This realization presupposes that our sense of sight *actually* extends to the objects seen. The proportionality of the depicted sizes of the objects on the retina cannot form a reference to the true sizes of the objects. We cannot simply assume that the brain develops the correct visual sizes from the image sizes on the retina by using a certain factor of magnification. This is because an object that is closer to the eye and relatively small, such as a table, appears on a larger surface of the retina than a distant but huge object, such as a mountain. The traditional and

[10] Thaliath, Babu (2001): *Der Prozeß der Geometrisch-Optischen Perspektivierung in der Visuellen Raumwahrnehmung*, FreiDok (Freiburger Dokumentenserver), Freiburg i.Br., http://www.freidok.uni-freiburg.de/volltexte/271, p. 37ff.

still current problem of the *object size constancy* in the visual size perception seems to prove that there is an *immediate optical connection* between the image on the retina and the actual size of the objects. In his work *Molyneux's Question. Vision, Touch and the Philosophy of Perception*, Michael M. Morgan discusses how Condillac sought to provide experimental proof of the aporia of the object size constancy and thus of the visual size perception. Condillac contradicts the prevailing doctrine of *unconscious inferences*, which was previously held by philosophers such as Locke and later by scientists of optics such as Helmholtz:

> Locke's error, as Condillac clearly points out, was to think that we see the retinal image at all. If we first see the flat image and then later perceive, Locke's argument (and Helmholtz's) follows: some process of inference must have go on. But if we never see the image – and Condillac correctly points out that we are never conscious of so doing – then the 'inference' is gratuitous. We do not and cannot see the retinal image: we see objects in the outside world. The Lockean and Helmholtzian language of 'unconscious inference' is an undesirable relic of the 'camera' theory of vision.
>
> In some respects Condillac thought more clearly about this problem than many contemporary psychologists. Take the question of 'object constancy' for example. Condillac knew that 'If a man four feet away ... steps backward to eight feet, the image of him on the retina is halved in size.' Because of this it has seemed even to some contemporary theorists to be a problem that objects do not shrink rapidly in size as they go away. Originally, the descriptive term 'object size constancy' was used to refer to the non-shrinkage phenomenon. Its use in that way is unexceptionable. But some people now use the term 'constancy' as if it applied to a process which set to work on the retinal image: they speak of constancy 'scaling things up' or 'scaling them down.' What exactly do they think is being altered in size by constancy? The size of objects? Obviously not. The retinal image? Still less so. The size of an image in the brain? Possibly: but for what purpose? A moment's thought shows the problems in treating constancy as a magnifying/minifying process. The cause of the fallacy is the belief that we see the retinal image.
>
> Condillac disposes of the fallacy. For one thing, he makes the very just remark that 'If perception is an inference involving a link between the idea of a man and a height of about five feet, either I should not see the man at all, or I should see him five feet tall' – whereas in fact objects seem to decrease insensibly in size as they move into the middle distance. He ends with the remark 'Nature determines that the sight of these objects should tell me how far the man is away; it is impossible that I should not have this impression every time I see them.' In other words, we see things as we do, not because we make inferences, but because we are as we are. As modern jargon would have it, the system is hard-wired.[11]

The "object size constancy" is an ophthalmic phenomenon that we experience constantly in our daily lives. If the visual size perception were based *solely* on the size of the image on the retina, the apparent size of objects in close proximity would change rapidly when they or the observer moved, increasing or decreasing in size. In the visual model shown above (Fig. 3.2), the reduction in the size of the retinal image is faster *according to the geometrical-optical principles* if the movement or the *moving away* of the visual object occurs in the near environment than at a later point in time when the visual object is far removed from the eye. That is, at greater distances, the reduction in the size of the retinal image becomes progressively smaller and

[11] Morgan, Michael J. (1977): *Molyneux's Question. Vision, Touch and the Philosophy of Perception*, Cambridge, pp. 78–79.

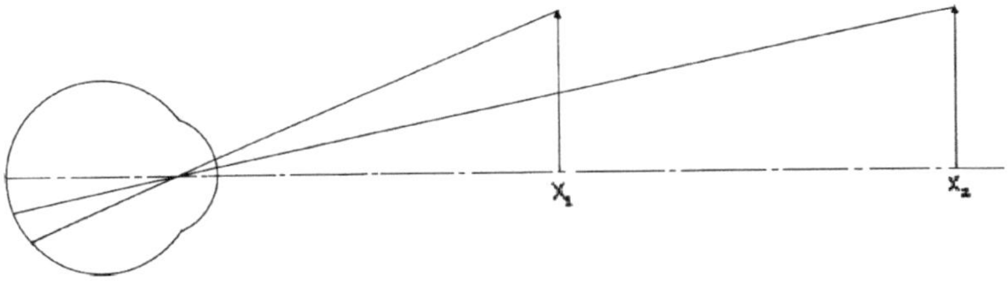

Fig. 3.2 The *object size constancy* in the visual size perception

slower; therefore, it is understandable that an object appears small at greater distances, when the size of the retinal image remains more or less constant and small compared to its initial rapid reduction. However, the phenomenon of *object size constancy*, in which the apparent size of the object directly seen remains unchanged despite the rapid and significant reduction in the size of the retinal image at the beginning (when the object is moving in the viewer's immediate vicinity), seems very puzzling to us. Since the rapid reduction of the retinal image theoretically contradicts the visually perceived *object size constancy*, this phenomenon seems to be based rather on the *size constancy* of the object itself. This supports the *object reference* in the visual size perception discussed above. In this way, the ophthalmic-optical phenomenon of *object size constancy* justifies the idea that the true reference in the visual size perception is the visual object itself. In the above-cited observation, Condillac refutes the thoroughly absurd assumption that the visual size of objects is dependent on their size on the retina for the following reasons (already explained):

1. We do not see the image on the retina, but the real objects in the visual field.
2. The visually perceived size of nearby objects remains almost unchanged, regardless of the rapid magnification or reduction of the retinal image.
3. The very small area of the retinal image cannot serve as a basis for the development of the approximately correct visual size perception of objects.

These considerations suggest that the actual act of seeing occurs in a unity of depth and size perception, in that the sense of sight takes on some of the characteristics of the sense of touch. Both in the visual perception of the size of nearby objects and in that of distant objects, we *experience* directly that visual perception of size arises in a *natural* connection with the perception of depth—which is most likely to be evidenced by the *visible* perspectivity of the visual image or field—so that we seem to be touching the objects visually, just as we can touch smaller and closer objects with our hands. We will discuss this unity between the visual depth and size perception in detail on the basis of the following reference points. It should be noted here that this unity, which can be experienced directly, ultimately seems to refer to the domain of *external* or extra-bodily geometrical optics. This geometric-perspectival structure of visual space is not *primarily* an internal construction—like imagination—but an external or extra-bodily experience. Now the question arises as to whether the direct experience of the unity between the perception of depth and size and all its essential features, such as the approximately correct sizes, positions and solidity of the nearby appearances, the perspectivity of the visual image, the visual perception of the free space (which is not depicted on the retina), etc., are actually developed solely in the brain by neuronal activity or whether they come about—geometrically and optically—in the *reality* of the extra-bodily visual space.

As discussed above, the image on the retina forms both the boundary and the link between the two parts of ophthalmic optics, i.e., between the geometrical-optical part and the physiological part, i.e., the part of the process of vision from the generation of photoelectric signals on the retina and their transmission to the brain through the optic nerves up to their neuronal processing. If the process of vision is limited to the physiological part alone, all the characteristics of the visual size perception discussed above should be generated or *caused* neuronally. The neuronal processes as the cause of the *object size constancy* discussed above could be demonstrated experimentally (although the neuronal basis of visual size perception is quite inadequate due to its ontological difference). That is, *object size constancy* in visual size perception could be supported by an equally consistent neural process base, even though the input—the size of the retinal image—changes radically. Nevertheless, it seems inconsistent to assume that the phenomenon of *object size constancy* is caused by neuronal causality alone. Rather, this causality seems to be *presupposed* by the actual geometric-perspectival structure of the visual space and the object reference in the visual size perception as described above. That is, the causality of *object size constancy* seems to come about *primarily* in the external geometrical-optical part of the visual process itself and *rather secondarily*, but inevitably, to condition the neuronal basis. In logical terms, it could be expressed as follows: for the visual phenomenon of the *object size constancy*, its neuronal process basis is a necessary, but not a sufficient, cause; it is, above all, a correlative or complementary causation. A clear distinction between a geometrical-optical effect and its purely neuronal causation is an undertaking that cannot be realized here. The causality of this and similar essential features of real vision seems to extend to both its purely physiological and its extra-bodily geometrical-optical processes. In short, in the case of visual size perception, a neuronal causality can hardly be separated from an extra-bodily and essentially geometric-optical sphere of effects or effectuation; rather, visual size perception seems to be based on the *operational* unity between the physiological-neuronal part and the extra-bodily geometric-optical part of the process of vision. The aporias of visual size perception discussed above can only be resolved by such a unity.

Just like static objects, movements of objects are also depicted in reverse and in a very diminutive way on the retina. As Fig. 3.1 shows, the movement of an object from A to B appears on the retina as the distance ab, which is a very tiny spatial distance compared to AB. According to this diminution of spatial distance, the imaged speed of the movement (from a to b) is extremely slow compared to the seen speed. How then is the real movement seen in approximately correct spatiotemporal extension, position or direction and speed with such an *input*? In the second case, in which the object approaches the eye along the visual axis, only a mere magnification of the object is imaged on the retina, and no spatial *extension* or distance of its movement. Nevertheless, this movement is visually perceived as it actually occurs in the real space. In both cases, we see how a movement that does not leave an appropriate image on the retina—as input—is perceived alone in direct vision. Therefore, the visual perception of the movement or dynamic objects—similar to the static objects—does not appear to arise solely through neural processes, but through the unity of the real visual space with the eye, as discussed above.

If we assume that the input for our visual perception of size is not only the image on the retina but also the light rays falling on the retina or converging on the eye point, i.e., the angle of their convergence, we tacitly assume that our sense of sight or our visual sensibility extends to an extra-bodily and purely physical medium such as light. But such an assumption—as a so-called "cue"—can hardly solve the problem of visual perception of size. This is because the convergence of light rays can only produce an approximately correct *geometric-optical* perception of size with the *actual distance* of the visual object. The distance of the visual object—as Berkeley emphasizes—cannot be depicted by the light rays on the retina. Furthermore, the inclusion of light rays in the process of vision implies the immeasurable extra-bodily expansion of vision in the real space, which contradicts the claim of physiological optics, according to which visual-spatial perception develops only neuronally.

3.2.3 Visual Perception of Position

Ever since the anatomical discovery of the retina in the eye, the geometric-optical inversion of the retinal image has been an aporia for which various solutions have been proposed in the history of the development of ophthalmic optics. All static phenomena and dynamic movements in the real visual space are imaged upside down and laterally reversed on the retina. Nevertheless, we see all phenomena and their movements upright and in the correct positions. The visual perception of the position is a mystery, because the images on the retina do not correspond to our physical upright existence and to the tactile perceptions of the objects. The idea that the correct visual perception of the position is suggested by the upright posture of the body and by the tactile perceptions seems hardly applicable here.[12] First, the optical and tactile perceptions of the objects point to completely different modes of sensibility; further, our bodily-tactile perceptions are limited to a few close objects, as discussed earlier. Huge and distant objects such as a mountain cannot be physically touched in their entirety, but only visually perceived.

The notion that the inverted retinal image is merely corrected neurally (which is assumed in the physiological optics) seems to be a naive belief. First of all, it does not solve the already described *problem of reference*. How does the brain *alone* determine that the objects and movements imaged on the retina are actually inverted and reversed? Again, the real objects and their real movements are the true, appropriate and apparently the only references for correct perception of position. Furthermore, any purely physiologically *discovered* neuronal causality, that *seems to solve* the

[12] Thaliath (2001, p. 73ff.).

problem of visual perception of position, is hardly ever proven by actual seeing alone. Since this *confirmation* cannot be found at the neuronal level, but rather at a completely emergent, extra-bodily and geometrical-optical level of effectuation, it can give the neuronal causality the necessity, but not the adequacy. That is to say, neuronal causality is necessary for the correct visual perception of position, but it is not sufficient. This is because the confirmation of the correct—or corrected—visual perception of position is limited to an extra-bodily geometric-optical *level of effect*, which is characterized by a complete ontological difference as against the physiological brain processes. It is therefore difficult to determine whether the causality of the correct visual perception of position is limited to neuronal processes alone or extends beyond them to the extra-bodily and geometrical-optical level of effect or effectuation itself.

The aporia of the visual spatial perception seems to suggest the previously discussed unity of the visual space with the eye, in which both the problem of the ontological difference between the physiological and the extra-bodily geometric-optical optics and the problem of the *reference* are suspended. If the subjective visual perception actually extends to the extra-bodily phenomenon of light and object space, the retinal image forms a mere link between the two domains of ocular optics discussed above. With regard to such a connection and the resulting unity of the visual space with the eye, the position of the phenomena imaged on the retina—or rather their inversion and lateral reversal—remains more or less *irrelevant* with regard to the correct visual perception of the position of the objects. The fact that objects are projected upside down onto the retina due to the geometric-optical functioning of the eye would therefore be a mystery programmed by nature to conceal an equally mysterious structure of vision.[13] However, in the modern developmental history of this discipline, when faced with the choice between the enigma of the inverted retinal image and the conviction of the perceptual-theoretical unity between the eye and the visual space, philosophers and scientists of ophthalmic optics seem to assume the former; thus, they tend to solve this enigma in the realm of physiological optics.

The invention of the *camera obscura* in the early modern period led to the eye-camera analogy in the field of ophthalmic optics. During this time, philosophers and scientists of optics had different stances over the inversion of the retinal image and its function in the process of vision. Many of them referred to an inevitable correction of this problem or of the inconsistency between the phenomenal reality and its retinal image. While some—in line with the eye-camera analogy—assumed a geometrical-optical correction, the others, especially the empiricists, believed in an empirical-perceptual development of the correct perception of position from the given inverted retinal image:

> For the empiricist philosophers there was another solution to the inverted image problem: experience. As we reach for objects we learn that we must reach upward for those that project images at the bottoms of our eyes and downward for those that project images at the tops of our eyes. This implies that up and down within the eye is initially noticed, and we learn to live with the inversion of the retinal image. Variations of this explanation, by Locke (1694),

[13] Ibid., p. 75.

> Berkeley (1709) and others, have greatly influenced experimental psychology. The inverted image issue has given rise to an area of research, perceptual adaptation, that is active and widespread today.[14]

The important experimental basis for the earlier investigations of "perceptual adaptation" was the reversal of the retinal image and its true function in the process of vision. In 1896, Stratton was the first to experimentally investigate visual adaptation using a geometrically and optically corrected, or upright, retinal image:

> The classic research in perceptual adaptation was reported by Stratton in 1896. Stratton was concerned with evaluating two theories that held the inversion of the retinal image to be necessary for the perception of objects as upright. According to the first theory, objects are projected back into space in the directions in which the rays of light fall upon the retina. The second theory holds that perception of upper or lower in the visual field depends on whether upward or downward eye movements are required to bring objects into focus. According to Stratton, both of these theories require that the retinal image be inverted in order that things be seen in an upright position. In order to test these theories, he developed an optical device that reinverted the image and wore this device for varying periods of time (Stratton, 1896, 1897a, b). He was able to adapt to the "upright" retinal image, at least to a degree. Stratton (1896) used this adaptation as evidence against theories that hold that an inverted retinal image is necessary for the perception of an upright world: 'In Fact, the difficulty of seeing things upright by means of upright retinal images seems to consist solely in the resistance offered by the long-established previous experience. ... But a person whose vision had from the very beginning been under the conditions we have in the present experiment artificially produced, could never possibly feel that such visual perceptions were inverted.[15]

After Stratton, analogous experiments on visual adaptation were conducted by many scientists. The results of these experiments were more or less consistent with Stratton's conclusions. The experimental evidence seems to defend the position of the empiricists over the nativists in terms of the development of the correct visual perception of position. The central question that arises with regard to the experimental evidence of the visual adaptation is: how does visual adaptation arise independently of the mode of retinal imaging? We cannot merely assume here that this otherness in the retinal imaging is corrected by an "adaptation." The blurring of the retinal image in the case of short- or long-sightedness, as well as the cylindrical distortion of the retinal image in the case of astigmatism, have a direct effect on the visual image and can only be corrected geometrical-optically by means of appropriate lenses. However, in case of the correct visual perception of position, the eye does not seem to seek a *reference* to the retinal image, but to the phenomena in the visual field itself. The visual adaptation according to the empirical explanation, in which the upright position of the objects as appearances in the immediate visual field is originally suggested by the familiar haptic perception of the upright positions of the objects *in terms of subjective theory of perception,* seems to be inconsistent here. If we venture into a hypothetical idea that all objects near and far (except the ground) that we cannot fully touch, such as trees, mountains, high-rise buildings, etc., are turned upside down, they will be depicted upright on the retina. However, we will always

---

[14] Braunstein (1976, p. 8).

[15] Ibid., pp. 8–9.

see them as they actually exist. Visual adaptation, as found in Stratton's experiment, cannot be the case here. This seems to indicate that the eye, when visually adapting to the correct perception of position, does not use the retinal image, but the position of the directly seen real object itself as a reference.

In short, the evidence of visual adaptation in the ophthalmic-optic experiments of Stratton and other scientists seems to confirm the previously discussed notion of the unity of the eye with the immediate visual space, where true references to the correct perceptions of size and position are the real objects themselves. For here the eye *adapts to the correct mode of appearance of the objects* regardless of their modes of representation on the retina. In other words, the retinal image has the essential function here of connecting the physiological part of the ocular optics with its geometrical-optical part and thereby creating the unity of the visual space with the eye, which *accordingly* is hardly affected by the geometrical-optically inverted or upright retinal images. That the eye *can visually adapt* to the correct position of the objects on the retina with both the naturally inverted and the geometrical-optically corrected or upright image demonstrates the fact that the basic reference for such optical adaptations is *the real object itself* and not its image on the retina. The empiricist position that this adaptation comes about solely on the basis of subjective theory of perception, and indeed suggested by the familiar haptic perception of objects, is untenable here. In Stratton's experiment, the objects initially appear upside down; the gradual inversion of the objects seen during the perceptive or visual adaptation is therefore a purely ocular phenomenon and not a mere subjective suggestion that is transferred to the immediate visual perception.

3.2.4 The Geometrical-Optical Structure of the Visual Space

The geometrical-optical structure of our immediate visual space is principally built on the light pyramid, which is constructed by the light rays reflected from the objects and converging on the eye point. The optical and dioptric qualities of the light rays—such as linearity and the geometric laws of the dioptric phenomena of reflection and refraction—give the structure of the visual space *natural geometric* characteristics, which are represented mainly in the perspectivity of direct vision and in the *geometric* laws of visual virtuality (reflection and refraction). How does the directly experienced geometric-optical structure of vision develop? The only input in the eye that results geometrical-optically from the light pyramid is the retinal image. But since the retinal image in the eye is *not seen* during the process of vision, the essential features of ordinary visibility of an image, such as colourfulness and perspectivity, cannot be attributed to the retinal image. Apart from colour, the entire three-dimensional perspectival structure of our immediate visual space should arise from a flat and *aperspective* retinal image.[16] According to Gibson, the development of visual depth

[16] The factum of stereoscopic vision discovered by Wheatstone can indeed generate visual depth perception (as demonstrated by stereoscopic virtuality), but it can only perfect an already existing

perception from a flat retinal image remains one of the unsolved mysteries of visual spatial perception for scientists in the field of ocular optics. Visual depth perception refers to some essential features of the visual space, such as its perspectival structure, the visual perception of the distance and solidity of objects, as well as the free or intermediate space that is not depicted on the retina. The problem or aporia of visual perception of size, depth and position is discussed in detail by George Berkeley in his first major work, *An Essay towards a New Theory of Vision.* Berkeley deals in principle with all geometric-optical attempts to explain the visual phenomenon—in particular with visual depth perception. Berkeley emphasizes that the lines and angles that, according to geometric-optical laws, should suggest depth perception, are neither perceived directly nor imaged on the retina:

> 1. My design is to shew the manner wherein we perceive by sight the distance, magnitude, and situation of objects. Also to consider the difference there is betwixt the ideas of sight and touch, and whether there be any idea common to both senses.
>
> 2. It is, I think, agreed by all that distance, of itself and immediately, cannot be seen. For distance being a line directed end-wise to the eye, it projects only one point in the fund of the eye, which point remains invariably the same, whether the distance be longer or shorter. [...]
>
> 11. Now from sect. 2 it is plain that distance is in its own nature imperceptible, and yet it is perceived by sight. It remains, therefore, that it be brought into view by means of some other idea that is itself immediately perceived in the act of vision.
>
> 12. But those lines and angles, by means whereof some men pretend to explain the perception of distance, are themselves not at all perceived, nor are they in truth ever thought of by those unskillful in optics. I appeal to everyone's experience whether upon sight of an object he computes its distance by the bigness of the angle made by the meeting of the two optic axes? Or whether he ever thinks of the greater or lesser divergency of the rays, which arrive from any point to his pupil? Everyone is himself the best judge of what he perceives, and what not. In vain shall any man tell me, that I perceive certain lines and angles which introduce into my mind the various ideas of distance, so long as I myself am conscious of no such thing.
>
> 13. Since, therefore, those angles and lines are not themselves perceived by sight, it follows from sect. 10 that the mind doth not by them judge of the distance of objects.
>
> 14. The truth of this assertion will be yet farther evident to anyone that considers those lines and angles have no real existence in nature, being only an hypothesis framed by the mathematicians, and by them introduced into optics, that they might treat of that science in a geometrical way.
>
> 15. The last reason I shall give for rejecting that doctrine is, that though we should grant the real existence of those optic angles, etc., and that it was possible for the mind to perceive them, yet these principles would not be found sufficient to explain the *phenomena* of distance, as shall be shewn hereafter.[17]

Both Berkeley and his opponents, who seek to defend the geometric-optical basis of visual perception, mainly refer to the same object of investigation, namely to the

---

depth perception—or rather the monocular depth perception—whose causes are to be sought elsewhere. See Thaliath (2001, pp. 80, 103–104). We discuss this in the course of our investigation.

[17] Berkeley, George (2009): *An Essay towards a New Theory of Vision*, Philosophical Works. ed. Michael R. Ayers, London, pp. 7–9.

light rays that perspectively structure the visual space and construct the retinal image in the eye. However, they do not start from the real or geometric-optical construction of the visual space by the light rays, but from its purely physiological-optical origin, which is *suggested* by the phenomena in the visual space or by the geometric-optical characteristics of the light rays converging on the eye point. Berkeley's arguments refer to the *invisibility* of the light rays; these imply that the geometrical-optical forms such as the lines or angles are not perceived by the eye. However, our real or directly experienced visual space is based on a *natural* geometrical-optical structure of the light rays reflected by the objects and converging on the eye point. All the essential traits of visual spatial perception discussed above, namely the size, distance and position of objects as *seen* and the extension of visual free space, seem to be incorporated in this single structure of the light pyramid. It seems that a *natural geometry* is inherent in the physical light rays and their construction or structuring of the light pyramid of our visual perception of space; it gives the sense of sight its real essential features, but also its virtuality, as represented in the dioptric phenomena such as reflection and refraction. While the perspectivity of the visual space can be derived more or less directly from this structure of light rays, it still needs to be examined in more detail how the above-mentioned and further essential features of visual spatial perception—including visual virtuality—are originally dependent on a natural geometry of light rays that construct the visual space and fill it.

In the early modern period, there were various geometrical-optical speculations regarding visual spatial perception—in particular, visual depth perception—against which Berkeley polemicized in his treatise. There were many attempts to explain the visual distance perception or trace it back *in terms of geometrical optics* to the angle of convergence of the light rays (reflected by objects) or the angle that the visual axes of the two eyes form with the object seen. There is no sufficient reason to assume that the eyes *calculate* the visual distance of the object from these given geometrical-optical premises. Similarly, the assumption that such geometrical-optical characteristics of vision *suggest* visual distance perception purely in the physiological-optical terms seems to be a belief. However, the perceptual inadequacy of such geometrical-optical premises in the study of the visual spatial perception does not exclude the natural geometry or geometrical optics that is briefly discussed above and their legitimacy in the process of vision. For our visual space, which encompasses objective phenomena and extension of free space, is *geometrically and perspectively structured* by the physical rays of light and their optical and dioptric properties.

According to Berkeley's polemic against the geometrical-optical interpretations of the process of vision, the geometric lines, angles, etc., are not *seen*. Equally invisible are the light rays that geometrical-optically structure the visual space; they are not *represented* on the retina—as Berkeley emphasizes—although they alone connect the real phenomena in the visual space with the retinal image in the eye. If—in line with Berkeley—the basic geometric forms such as line and angle, which are the *natural-geometric* characteristics of the light rays converging on the eye point, cannot, strictly speaking, be *any inputs* in the process of vision, how does the geometrical-optical structure of our visual space actually arise in approximate accordance with the geometrical-optical and perspectival structure of the *real* light rays? Here we cannot trace the geometrical-optical structure of our immediate visual space back to the retinal image, *if* it is the only *input* in the process of vision. For neither the depth of the visual space nor its perspectivity is imaged on the retina. The invisible flat image on the retina does have a perspectival structure—like the image in a camera—but it cannot claim perspectivity without the factum of direct spatial vision. For perspectivity—that is, the *virtual* perception of depth in a flat image—arises only in the direct *spatial vision*. As already discussed, the geometrical-optical structure of our visual space can be recognized by the perspectivity, dioptric virtuality, or the directionality of vision.

3.2.5 *The Perspectivity of Vision*

The discovery of central perspective during the Renaissance by Filippo Brunelleschi was, as is well known, a geometrical process, the basis of which was the convergence of the orthogonal lines to the vanishing point. This convergence of the geometric orthogonal lines obviously represented the real convergence of the light rays to the eye point in the actual process of vision. Likewise, the single vanishing point[18] in the geometrical-optical construct represented the single eye point. That is, the correct procedure of the geometrical perspectival construction was developed by the civil engineer Brunelleschi directly from the structure of the light pyramid during the act of seeing. This traditional method of geometrical perspective is still used in painting and especially in architectural drawings, and as such forms an important propaedeutic in these fields. In his essay "Perspektive als Symbolische Form" (Perspective as Symbolic Form), Erwin Panofsky explains the conventional method of the geometric technique of central perspective:

> This correct construction was in fact invented in the Renaissance, and although later subjected to various technical improvements and simplifications, it nevertheless remained in its premises and goals unchanged to the time of Desargues. It is most simply explained as follows: I imagine the picture — in accord with the "window" definition — as a planar cross section through the so-called visual pyramid; the apex of this pyramid is the eye, which is then connected with individual points within the space to be represented. Because the relative position of these "visual rays" determines the apparent position of the corresponding points in the visual image, I need only draw the entire system in plan and elevation in order to determine the figure appearing on the intersecting surface. The plan yields the width, the elevation yields the height; and if I combine these values on a third drawing, I will obtain the desired perspectival projection (Figure 1).[19]

---

[18] The discovery of the central perspective with a single vanishing point has a history of development in the Middle Ages. In the earlier attempts at perspectival representations by medieval painters such as Giotto, Duccio and Masaccio, the orthogonals converged on several vanishing points. Brunelleschi's discovery of a single vanishing point led to the correct geometrical method of central perspective, which is still used in painting and engineering, especially in architecture. See Thaliath (2005, pp. 201–203, 327 f.).

[19] See Fig. 3.3, see Panofsky, Erwin (1991): *Perspective as Symbolic Form*, trans. Christopher S. Wood, Zone Books, New York, p. 29.

> In a picture constructed this way – that is, by means of what Dürer called a "planar, transparent intersection of all those rays that fall from the eye onto the object it sees" – the following laws are valid. First, all perpendiculars or "orthogonals" meet at the so-called central vanishing point, which is determined by the perpendicular drawn from the eye to the picture plane. Second, all parallels, in whichever direction they lie, have a common vanishing point. If they lie in a horizontal plane, then their vanishing point lies always on the so-called horizon, that is, on the horizontal line through the central vanishing point. If, moreover, they happen to form a 45-degree angle with the picture plane, then the distance between their vanishing point and the central vanishing point is equal to the distance between the eye and the picture plane. Finally, equal dimensions diminish progressively as they recede in space, so that any portion of the picture – assuming that the location of eye is known – is calculable from the preceding or following portion.[20]

As Panofsky describes it, the geometrical method of central perspective was originally developed from the window definition of the image, which was a prevailing idea in the Renaissance. If the perspectival image is a cross-section of the visual pyramid (shown in Fig. 3.3), this visual pyramid arises from a real physical light pyramid. The orthogonals in the geometric process of central perspective, although they determine the *seen* position and size of the phenomena as "visual rays", represent, in geometrical-optical terms, the physical light rays themselves, which are reflected by the objects and converge on the eye point. This is clearly demonstrated in Dürer's endeavor to develop the construction of geometrical central perspective directly from the empirical vision (see Fig. 3.4).

The importance or basis of the orthogonals in the geometrical construction of central perspective indicates that this process can originally be developed from the structure of the visual space itself, which the physical light pyramid generates. In real vision, the orthogonals *in conformity* with the physical rays of light form the part, or more precisely the corpus of geometrical optics, which gives rise to the image on the retina. If we dispense with this corpus of the light pyramid in our interpretation of the perspectivity of seeing, only the retinal image remains as the only input in the process of vision. Now the entire *structure of* perspectival vision must be developed solely from this input in the eye. Such an assumption seems to be inconsistent from the outset for the following reasons:

1. The problem discussed earlier, that in the process of seeing we can only *see* real objects in the visual field and not their images on the retina, is of great importance here. This is because the perspectivity of a two-dimensional image primarily requires *perspectival* vision. The *unseen* retinal image is, as already explained, a mosaic of different light zones, shadowy boundary lines, etc., and therefore has no perspectivity. If we assume that the perspectivity of the visual space that is directly experienced in vision is originally developed from the retinal image—based on the eye-camera-analogy—we tacitly ascribe perspectivity to the retinal image and thus a certain intraocular visibility (against which philosophers and scientists of optics such as Descartes, Kepler and others have polemicized). The retinal image is not seen in the process of vision, but it forms a link between the physiological and the

[20] Ibid., pp. 27–28. Also see Panofsky, Erwin (1998): *Die Perspektive als symbolische Form*, in: *Deutschsprachige Aufsätze II*, eds. Karen Michels und Martin Warnke, Akademie Verlag, Berlin, pp. 665–666.

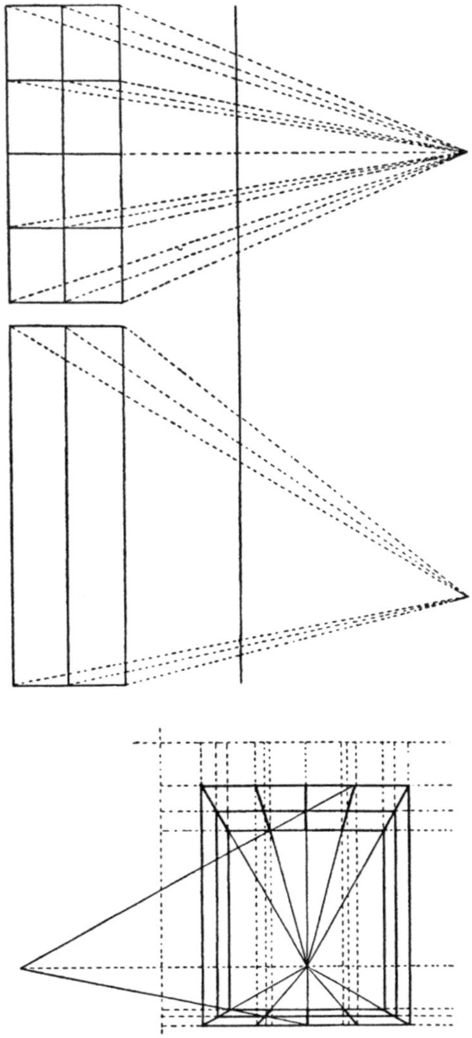

Fig. 3.3 Modern "linear perspectival" construction of a rectangular interior space ("spacebox"). Top: plan. Middle: elevation. Bottom: perspectival image arrived at by combining the segments marked off on the "projection line". Panofsky, Erwin (1991): *Perspective as Symbolic Form*, trans. Christopher S. Wood, Zone Books, New York, p. 29

geometric-optical part of the vision. Such a link and the resulting unity of the visual space with the eye would also suggest that this visually perceptible perspective or perspectival structure actually comes about in the real visual space—and indeed geometric-optically through the light rays.

2. The perspectivity of seeing, to which the visual depth perception in particular refers, seems to have been originally developed from an *empirical* familiarization with the real orthogonal lines representing the rays of light that converge on eye point in the real visual space. Without such empirical familiarization, we are unable to visually perceive either the *free spatial* depth of the depicted objects or their solidity in a *given* perspectival image. This fact was proved for the first time by

Fig. 3.4 Albrecht Dürer: *Draughtsman Making a Perspective Drawing of a Reclining Woman.*[21] Breidbach, Angela (2017): *Übertragene Körper. Diskurse des Schattens im Werk von Hans-Peter Feldmann, W. G. Sebald und William Kentridge*, Verlag Wilhelm Fink, Paderborn, p. 20

the case of Cheselden's patient. In the seventeenth century, the surgeon Cheselden was able to restore the sight of a 14-year-old boy who had been blind from birth by performing cataract surgery at St. Thomas's Hospital in London. Cheselden reports how his patient was only able to get used to perspective vision empirically:

> When he first saw, he was so far from making any judgement about distances, that he thought all objects whatever touch'd his eyes (as he expressed it), as what he felt, did his skin; and thought no objects so agreeable as those which were smooth and regular, tho' he could form no judgement of their shape, or guess what it was in any object that was pleasing to him: he knew not the shape of anything, nor any one thing from another, however different in shape, or magnitude; but upon being told what things were, whose form he before knew from feeling, he would carefully observe, that he might know them again; but having too many objects to learn at once, he forgot many of them; and (as he said) at first he learn'd to know, and again forgot a thousand things in a day. One particular only (tho' it may appear trifling) I will relate; having often forgot which was the cat, and which the dog, he was asham'd to ask; but catching the cat (which he knew by feeling) he was observ'd to look at her steadfastly, and letting her down again, said, So Puss! I shall know you another time. He was very much surpriz'd, that those things which he had lik'd best, did not appear most agreeable to his eyes, expecting those persons would appear most beautiful that he lov'd most, and such things to be most agreeable to his sight that were so his taste. We thought he soon knew what pictures represented, which were shew'd to him, but we found afterwards we were mistaken; for about two months after he was couch'd, he discovered at once, they represented solid bodies; when to that time he considered them only as partly-colour'd planes, or surfaces diversified with variety of paint; but even then he was no less surpriz'd, expecting the pictures would feel like the things they represented, and was amaz'd when he found those parts, which by their light and shadow appear'd now round and uneven, felt only flat like the rest; and asked which was the lying sense, feeling or seeing?[22]

The question: *which was the lying sense, feeling or seeing?* also explains the necessary development of the perspectival depth perception in direct vision, from which

[21] *Draughtsman Making a Perspective Drawing of a Reclining Woman* by Albrecht Dürer. See Breidbach, Angela (2017): *Übertragene Körper. Diskurse des Schattens im Werk von Hans-Peter Feldmann, W. G. Sebald und William Kentridge*, Verlag Wilhelm Fink, Paderborn, p. 20.

[22] Morgan (1977, pp. 17–21).

eye's ability to experience the virtual perspective in an image arises. This familiarization with perspective is obviously a familiarization with the *real* orthogonals or those penetrating into the depth of the visual space, which are constructed by the rays of light converging at the eye point—in a structural unity.

3. The perspectivity of our immediate visual space differs from the *virtual* perspectivity of a picture precisely in the perfection of the depth perception, which includes the extension of the free space, the free-spatial distances and the solidity of the objects. Even a picture that is completely perspectival can at most evoke a perfect or highly developed perspectival virtuality, whereas the perspectival nature of our direct visual space is not a virtuality but a *real,* visual experience, with which the virtual perspectival nature of a picture cannot be equated.[23] How does the *real* perspectivity of the visual space develop, even when the flat retinal image only resembles a perspectival image *in structure*? If real perspectival vision is developed solely from the retinal image and its perspectivity, then a perspectival image—a painting or a photograph—*should be able* to evoke an almost perfect virtuality of visual-perspectival depth perception. That is, we should be able to look at a perspectival image as if we were looking through a window into the depth of a real visual space.

However, a perfectly perspectival image—a Renaissance painting, a photograph or a cinematic image—can at best create a perfect perspectival virtuality of the perception of depth and solidity, but it cannot guarantee spatial vision. Even if such a perspectival image creates or reconstructs *the same* retinal image during direct spatial vision, the eye can only develop a perfectly perspectival virtual depth from this image, and no real or spatial vision. If the perspectivity of the real vision is developed solely from the retinal image, then an imperfect or approximately perfect perspectival image should also evoke the virtuality of real vision—as in a reflection. For the eye is, as demonstrated in the everyday phenomenon of reflection or in technically advanced stereoscopy, subject to the visual virtuality.

The fact that a perspectival image, even if it creates the same retinal image (as in real vision) in the eye, can only produce a perspectival virtuality and not a real spatial visual experience, also seems to be supported by the finding discussed above that real vision requires a real visual space.[24] This visual space is said to have developed from the *orthogonal* beams of light reflected by the objects, which penetrate the equally real free space and converge on the eye point, and to be structured by these.

4. In addition to their depth and solidity perception, the size and position of the objects seen, as described above, seems to be *merely constructed* directly through the perspectivity of seeing. In the case of geometrized visual objects, such as architectural structures, furniture, mechanical or electrical devices, etc., the perspectival construction is *very visible* in direct vision. You can immediately *see* how the objects

[23] We will not consider the possibility of stereoscopy here, i.e. creating a virtual sense of depth from a flat image based on the principle of binocular vision, since the perspectival perception of depth does not necessarily require binocular or stereoscopic vision. Both the virtual perspective of an image and the real perspective of the visual space can be perceived with one eye, albeit imperfectly.

[24] Thaliath (2005, pp. 80–81).

appear in the visual space *perspectively constructed* by having the clear impression that the essential features of the objective appearance, such as size, position, spatial distance and, above all, the perspectively-deformed extension of the objects, are more or less geometrically constructed by the orthogonal lines *projected* by the eye onto the objects. The actual sizes and correct positions of the objects are not depicted on the retina, nor do the free spatial distances and the solidity of the objects *appear*. It is not conclusive to assume that the two-dimensional structure of the retinal image suggests the correct spatial vision—with a spatial-perspectival structure and approximately correct perception of size, distance and solidity. Rather, the *actual perspectival* structure of the visual field seems to be constructed by *actual orthogonals*.

When we stand in front of many relatively close geometric or geometricized objects—as in a modern residential area—we see all the objects structured or deformed in the geometrical perspective. Let us imagine that the real perspectival structure of our visual space with the orthogonals converging at the vanishing point is constructed or sculpted objectively, i.e., geometrically and architecturally. Such a *life-size* representation consists—except for the *actual real* objects—of the of orthogonals that converge at the vanishing point, which is the *reversal* of the viewer's eye point, and are, as such, analogous to the light rays that are reflected by the objects and penetrate the free space and converge at the eye point. This *sculptural* representation shows that the observer sees the objects in the same perspectival structure—with approximately the correct perspectival size, distance, solidity and position—as they are actually constructed in perspective by the orthogonal lines, and that the directly seen perspectival deformations arise in their proper measure from the real pyramidal structure of the orthogonal lines.

The reality of the structure of the orthogonal in the immediate process of vision presupposes that our sense of sight extends beyond the body either through the orthogonally structured rays of light or through the visual rays projectively emitted by the eye, which form a structural identity with the converging rays of light. In such an extension of the sense of sight, the objects in the immediate visual field are *optically* touched. That is, the observer experiences the nearby objects in the visual field in approximately the correct size, distance, solidity and position and in an overall perspectival structure through an extra-bodily *optical* sense of touch. The *optical* orthogonals in the visual space seem to take on the function of two linear sticks crossing at one point, which the blind man in the Cartesian parable uses to perceive the extra-bodily objects haptically.[25] Now, the question arises as to whether the reality or materiality of the orthogonal lines that structure our immediate visual space in perspective is *analogous* to or even identical to the reality of the light rays converging at the eye point. The visual virtuality, represented in reflection and refraction, seems to be constructed *geometrical-optically* by the virtual orthogonal lines. The visual virtuality of reflection and refraction seems to prove *prima facie* the virtuality or purely the subjective origin of the orthogonal in vision, but in fact it *supports* the

[25] See Thaliath, Babu (2008): *Die Räumlichkeit der Empfindung*, FreiDok, Universität Freiburg, Freiburg i. Br., http://www.freidok.uni-freiburg.de/volltexte/4937/, pp. 19–21.

reality of the optical orthogonal discussed above, which enables the viewer to see properly—with the correct perception of the size, distance and position of the objects and their perspectivity. The distinction between optical reality and virtuality can be traced back to the constellation of structural unity and the ontological difference between the rays of light and the visual rays.

3.2.6 Optical Virtuality

The forms of optical virtuality, such as the dioptric phenomena (reflection and refraction), the perspectival depth perception in an image or purely mental phenomena such as imagination or dreams, differ in nature and degree of perfection. Among the various forms of optical virtuality, dioptric reflection and refraction constitute the clearest and most perfect virtuality, because they are almost identical to direct vision. All forms of optical virtuality—sensible virtuality in general—have an important essential feature in common, namely the *absence* of real objects. These forms of virtuality differ from the *real,* actual seeing; in them the objects *seen* are not real, but virtual, i.e. they *do not exist*. In case of the dioptric phenomena, virtual appearances arise solely from the virtual perceptions of the size, distance and position of the objects, which are, however, *present* in this process of vision, whereas in the case of fantasies, in the perspectival perception of images and in dreams, real objects are completely absent.

The forms of optical virtuality seem to support the generally accepted notion that seeing is a purely mental or *a priori* subjective construct. This is because the subject develops visual virtuality without the *direct* involvement of real objects. However, if we insist on such an assumption, we are inevitably confronted with a problem in the theory of perception that raises several questions: How does the *real*, actual seeing come about, in which objects are directly involved, and which as such can be distinguished from the dioptric virtuality such as reflection and refraction, but not excluded by them? How do real objects *enter* the mind as visual phenomena—with the correct size, distance, position and solidity, but also with the correct perspectivity—that the mind has *at its disposal* when it *generates* optical virtuality, such as in fantasy or dreams? How does the geometrical-optical structure of the dioptric reflection and refraction develop, in which the size, distance and position of the virtual phenomena perceived visually can be objectively or *geometrical-optically calculated or determined*? etc. All these problems or questions indicate that optical virtuality of this kind originally presupposes the actual seeing of objects.

We are referring in particular to the dioptric virtuality such as reflection and refraction (because they arise in direct seeing, in which real objects participate), so that we can clearly separate the optical virtuality from the optical reality and examine it. The virtuality of reflection manifests itself in many natural and technical phenomena—for example, in all types of natural reflection on smooth surfaces, on the surface of water, etc., as well as in artificial reflections through plane, concave and convex mirrors. Likewise, the virtuality of refraction manifests itself in the appearances in

Fig. 3.5 Reflection in a plane mirror

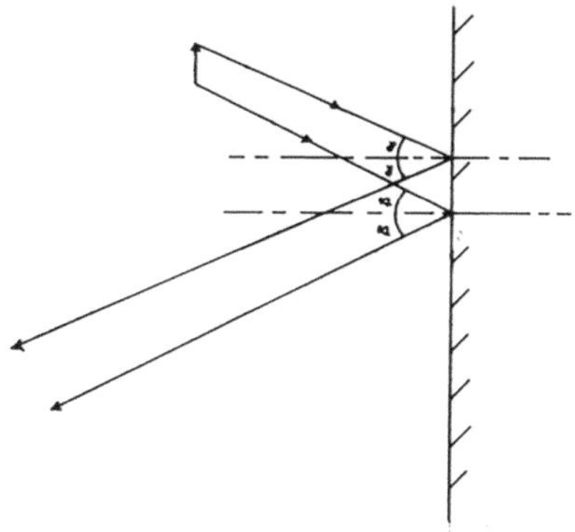

water or in a *fata morgana*, as well as in all forms of refraction through lenses and prisms in optical devices—from a simple magnifying glass to more complex optical devices such as cameras, microscopes and telescopes. All these natural and technical modes of dioptric virtuality have one important characteristic in common: *they arise in more or less exact geometrical-optical regularity, which can be determined objectively or calculated.* This means that the eye sees the virtual phenomena in an almost exact geometrical-optical proportionality of size, distance and position, and in an all-encompassing perspectival structure, as it can be objectively calculated or geometrically or even architecturally constructed. In short, all forms of dioptric virtuality are based on geometrical-optical laws, which seems to be the best way to prove their *objectivity*. The following figures represent a few simple and elementary modes of reflection and refraction:

Figures 3.5 and 3.6 show the reflection of light rays by plane and concave mirrors, and Figs. 3.7 and 3.8 show their refraction through a convex lens and a prism.[26] The straight lines show the direction of the light rays that are reflected and refracted. These purely physical properties of light, namely reflection and refraction, follow the laws of geometrical optics. The light rays in Fig. 3.5, which fall from an object onto a plane mirror, form the same reflection angles (α and β) when reflected; the light rays in Fig. 3.6, which come from an object located between a concave mirror and its focus (the midpoint of the radius of curvature) and fall on the concave mirror, are reflected according to different *geometrical-optical laws*. The reflection of the ray (1), which falls on the mirror parallel to axis XX', passes through the focus; the ray (2), which is geometrically represented as an extension of the radius (of the mirror), forms a single entity with the reflected ray. The light beam (3), which, if extended linearly, passes through the focus, is reflected parallel to the axis XX', and the light beam (4), which falls on the center of the mirror, forms the same reflection angles in its reflection.

All these representations, which are based on the geometrical-optical laws of the dioptric phenomena of light—such as reflection or refraction—are, strictly speaking,

[26] The following illustrations of reflection and refraction and their explanations are taken from my dissertation, which was published in 2005. See Thaliath (2005, pp. 205–208).

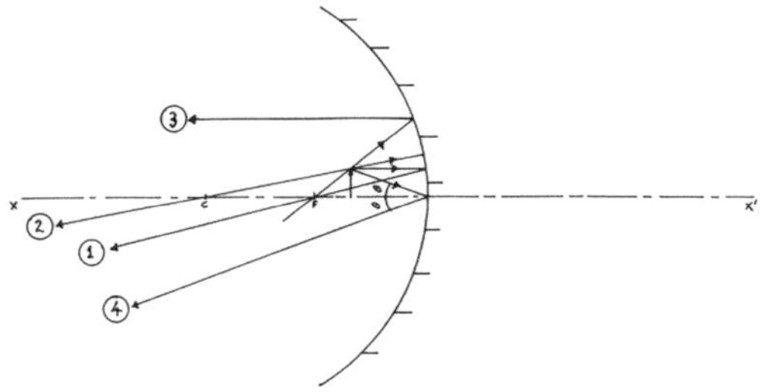

Fig. 3.6 Reflection in a concave mirror

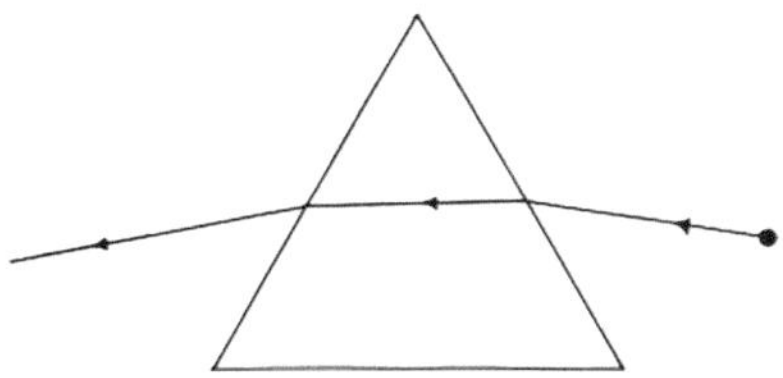

Fig. 3.7 Refraction through a prism

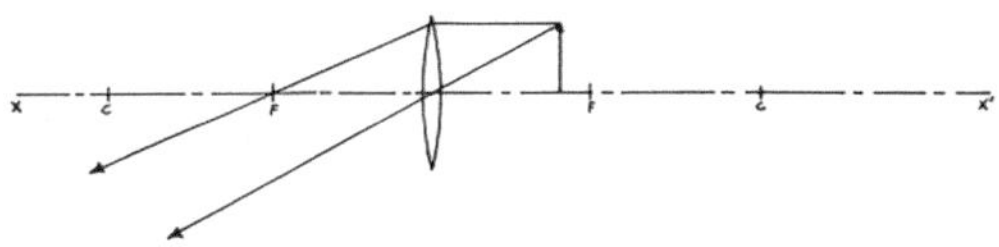

Fig. 3.8 Refraction through a convex lens

constructions of a theoretical space, which as such are *objectively* real. That is, they incorporate the real phenomena in space and the equally real light that fills the space. But when the eye or the factum of direct vision is included in these *real* designs, virtualities arise, a virtual space arises, which is essentially an aesthetic space, as the *modified* representations (Figs. 3.9, 3.10, 3.11 and 3.12) show. Figures 3.9 and 3.10 show the reflection of a *real,* material object; the *virtual* reflection appears behind the plane of the mirror. Similarly, the *view through* a prism or a convex lens shows the virtual images of a real object. The virtuality that comes about in these representations through the inclusion of the eye *occurs* in space, or more precisely in the aesthetic space of vision; the virtual images continue to be *constructed* geometrical-optically. What would be the basic principle of this construction of virtuality, which is generally evident in all these representations?

The characteristics of the virtual appearances, especially the perceptions of size and distance, in the forms of dioptric virtuality presented above are determined geometrical-optically and seem to arise as such *in the actual visual space.* That is to say, the *structural* causality of the dioptric virtuality, which can be calculated exactly in geometric-optical terms, appears to be the geometric-optical laws of the immediate visual space itself. These laws rely on the *natural geometry* of light rays and not only on the neuronal processing of the tiny retinal image in the context of physiological optics. If the dioptric virtuality does not arise in the immediate visual space, it is supposed to be constructed solely by the neuronal processes that

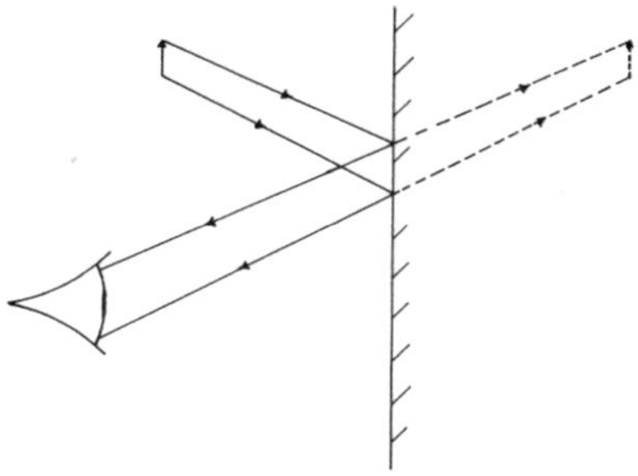

Fig. 3.9 Reflection in a plane mirror (in direct vision)

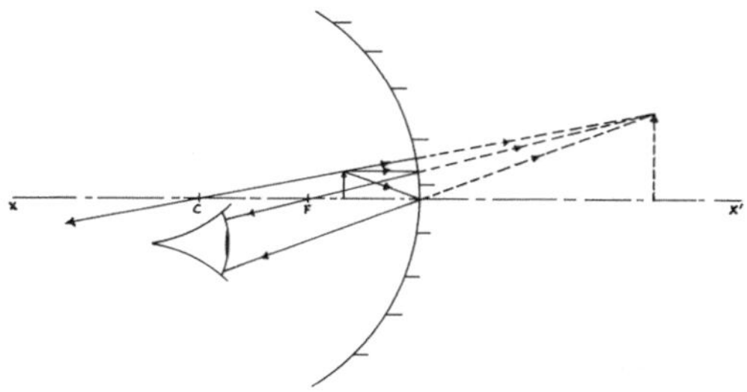

Fig. 3.10 Reflection in a concave mirror (in direct vision)

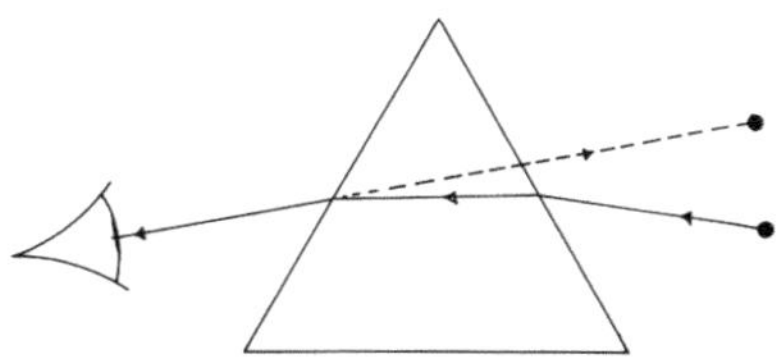

Fig. 3.11 Refraction through a prism (in direct vision)

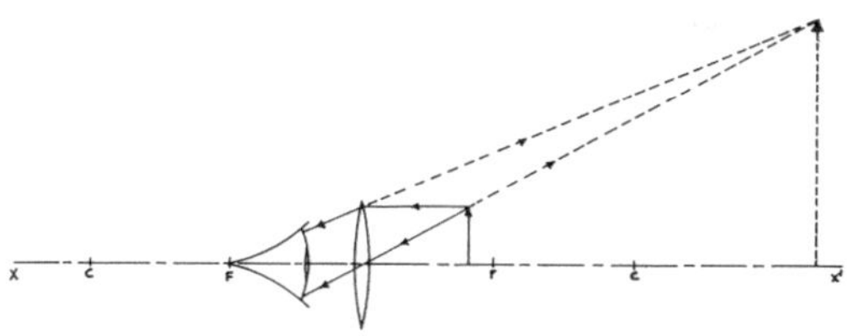

Fig. 3.12 Refraction through a convex lens (in direct vision)

only have the retinal image as input. This means that the brain should determine the geometrical-optical laws of the virtuality and accordingly develop the visual virtuality—alongside the real vision—only mentally. This seems very unlikely. For the *input* on the retina is quite insufficient to *construct* the above-described and other similar forms of dioptric virtuality in exact geometrical-optical lawfulness or regularity. When the (virtual) appearance is magnified by reflection in a concave mirror and refraction through a convex lens (shown in Figs. 3.10 and 3.12), the eye only has an enlarged retinal image of the visual object as input. Even if we claim that the magnification of the virtual appearances in these cases comes about through the magnification of the retinal image, we can hardly explain how the perception of distance or the visual perception of the exact spatial localization of the virtual appearances is developed solely from the retinal image. Obviously, the light rays reflected by the mirror or deflected by the lens and falling on the eye are to be *further followed* in order to *see* the geometrical-optically constructed virtual phenomena in approximately correct size and distance.[27] This would mean that real process of vision extends beyond the retinal image to the physical light rays and further to certain visual rays that *extend* the geometric-optical structure of the light rays into *virtuality*. This is because the *geometric lines* that construct the optical virtuality in the dioptric models of vision presented above—in accordance with the laws of geometric optics—are *merely linear extensions* of the light rays beyond the plane of the mirror or lens. By forming a unified perspectival structure of the visual space together with the real light rays, which includes both the real and the virtual appearances, the dioptric phenomena of reflection and refraction give the impression that these geometrical lines as *visual rays*, which in their geometrical-optical structure form a *unity* with the physical light rays and *extend* their linearity while *constructing* the optical virtuality, can claim a physical or objective-material phenomenality just like the physical light rays. Are they certain electromagnetic rays, which, unlike light rays, are not subject to reflection and refraction and therefore *separate themselves* from the light rays—i.e., from the unity with the light rays—during these dioptric phenomena, generating thereby the optical virtuality in accordance with strict geometrical-optical laws?

In the discourse on dioptric virtuality (on the basis of which we originally seek to prove the purely physiological-optical development of vision), the *actual* extension of vision into a real visual space would appear to us as a necessary factum. If we deny this objective factum and insist on the purely neuronal development of vision, the problem of the complete emergence of mental states, as discussed earlier, will cause us great difficulties. For here we are tacitly assuming that the brain, based on the input of the retinal image alone, computes all the essential features of the actual

[27] The plane mirror reflects the image realistically—that is, without changing its size and at the *correct distance* (from the mirror), just as we see objects with our naked eyes. Only the positions of the objects appear laterally reversed *according to the geometrical-optical principles*. The reflection on a plane mirror could *prima facie* be accepted as a striking example of how the eye develops a virtual visual image from the retinal image, which is completely similar to the real visual image (which includes the real objects). In a mirror, however, we do *see* the objects in their *correct* sizes and distances, as if they are *objectively* constructed corresponding to the geometrical-optical principles, that is, as if they are constructed *in reality* (as shown in Fig. 3.9). It is understandable that the real objects appear virtually in the same size on a plane mirror. This is because when reflected by a plane mirror, the size of the objects or appearances imaged on the retina remains unchanged—i.e., neither enlarged nor reduced. But the distances between the objects (which, strictly speaking, are not imaged on the retina) and the free spatial expansion 'seen' in the plane mirror image do not appear to be developed from the retinal image alone, but rather to be *geometrical-optically constructed* in the actual visual space itself. This is because we *see* objects—especially close objects—at the same distance on the plane mirror, which results from an objective *geometrical-optical construction* based on the optical and dioptric properties of light rays. This also applies to reflections in a concave or convex mirror.

spatial visual image *with precision* and *applies them* in an immediate visual experience. Such an assumption seems implausible from the outset. Firstly, here there is a completely ontological leap from a purely material to a mental phenomenon; the neuronal causality of the mental effectuation of the real visual image does not include any properties comparable to the optical and geometrical-optical characteristics of the real visual image. Therefore, it is almost impossible to explain the real visual image in terms of the theory of perception solely on the basis of material and neuronal causes. Secondly, it is equally inappropriate to assume that the geometrical-optical and perspectival structure of the visual space, which is already *objectively present* through the physical light rays, and whose image on the retina is quite inadequate, is *again* merely developed neuronally and projected onto the same already existing structure of the visual space. It seems much more plausible to assume that the geometrical-optical perception of size, distance, position and solidity—both in real and in virtual phenomena—and the entire perspectival structure of the visual space arise from an *original unity* of the eye with a real or objective visual space, whereby the true causality of the geometrical-optical structure of the visual space is not to be found in the brain, but *primarily* in the already existing physical-optical structure of the visual space and its geometric-optical and dioptric laws.

Berkeley's criticism of the legitimacy of the geometrical-optical interpretations with regard to the visual perception of size, distance and position can be turned around in view of the points discussed above. The lines and angles that form the geometric elements of the perspectival structure of the immediate visual space are neither seen nor imaged on the retina. But they arise and exist primarily through the physical rays of light and apparently also through the visual rays, which together form the real structure of the visual space. Light rays as elementary-structural components of the perspectival visual space are invisible *in themselves*, but they can be recognized in all the essential features of the visual space (such as the perception of size, distance, position and solidity of objects, the optical perception of free space and the overall perspectival structure of the visual space). The real extension of vision and its perspectival structure—which can be recognized first and foremost by the perspectival deformation of real and virtual appearances—are constructed primarily by light rays and their geometrical-optical *properties*. Berkeley *justifiably* polemicizes against the merely geometrical-optical explanations and justifications of the visual perception of distance, size and position by rejecting the assumption that the mind calculates the distance of the objects, their sizes, etc., through the given geometric lines and angles in the structure of the visual space. But the geometric-optical basis of the visual spatial perception is perfectly legitimate if the original or constructed perspectival structure of the visual space, which includes the real objects and the equally real free space, is taken into account. In case of the optical virtuality discussed above, this fact seems rather plausible to us. This is because the geometrical-optical legitimacy of dioptric virtuality can be recognized *primarily* by the geometrical-optical structure of the visual space itself and can hardly be *derived* from the entirely inadequate retinal image.

Berkeley rightly disputes the assumption that mind can perform the geometric-optical calculations with geometrical forms such as lines and angles—even if they are *given* in the visual process—and thereby merely construct the visual space. But our visual space, with both real and virtual phenomena, arises with approximately exact geometrical-optical form and conformity to the laws of the physical light-space that connects the eye with the objects of vision. From this it can be concluded that the geometrical-optical structure of the visual space that can be experienced directly is more likely to be present and given as such and not merely developed neurally. The natural geometry of light rays, which in *actual* vision can be traced back to their essential features such as linearity and *orthogonal-perspectival* convergence, gives rise to such a given perspectival visual space. In case of the forms of dioptric virtuality (discussed above), the *real givenness* of the geometrical-optical visual space should be clearer to us, because these forms of virtuality arise as *geometrical-optical extensions* of the *given* light rays through certain linear visual rays that are not subject to dioptric phenomena such as reflection and refraction, as any representation of reflection or refraction shows. In the case of dioptric phenomena, it seems that the visual rays, which form a unity with the light rays and thus give rise to the real visual space, are separated in the case of dioptric virtuality from the light rays—when reflected or deflected during refraction—and thus *geometrical-optically construct* the optical virtuality.[28] In the case of both real vision and virtual dioptric phenomena, it is inappropriate to assume that the brain recalculates the geometrical-optical structure of the visual space, which is *already given*, and produces it *merely subjectively* on the basis of the image on the retina and a few "cues" that do not form *sufficient* inputs during the visual process. For the neuronal processes that seem to *support* the geometric-optical characteristics of the visual space, such as the perception of size, distance and solidity as well as perspectivity, are in a fundamentally different mode of being as purely material phenomena; the causality of the geometrical-optical visual space (which comes about at a level of the effectuation of seeing) cannot therefore be merely reduced to the brain, i.e., to the purely physiological part of the optics of vision, but extends—especially with regard to the geometrical-optical and perspectival structure of the visual space—to the *given structure* of the visual light-space itself, as already discussed. In this case, the neuronal causality of vision is not to be imagined as the sole and sufficient *causa efficiente* and as such completely isolated from the real visual space, but rather as *participatory* and correlative, and indeed synthetic, i.e., in a synthetic nexus with the geometrical-optical laws of a real and objectively given light space. The fact that our sense of sight *necessarily has to extend to* an objectively given and perspectively structured light space seems to be sufficiently proven by the direct visual perception of the free space between objects.

[28] See Thaliath (2005, pp. 206–208).

3.2.7 *The Visual Perception of Free Space*

To begin with, we will refer back to the propaedeutic emphasis of Berkeley in his major work "An Essay Towards a New Theory of Vision", that has already been cited:

> It is, I think, agreed by all that distance, of itself and immediately, cannot be seen. For distance being a line directed end-wise to the eye, it projects only one point in the fund of the eye, which point remains invariably the same, whether the distance be longer or shorter.[29]

Although Berkeley hardly explains this in detail, in this consideration he refers to an important component of our immediate visual space, i.e., the free space between objects that the light rays reflected by material objects penetrate when they converge at the eye point. Likewise, his observation "distance being a line directed end-wise to the eye" refers to the light rays, which, due to their orthogonal structure, are only imaged on the retina as certain *points* during the process of vision. Both these facts form a structural unity in seeing, because the visual free space is created by the *invisible* light rays that penetrate it and give it fullness. Visual free space differs from visual objects in that it is not *seen*, but is perceived visually—as a purely extended nothing. Without this direct perception of the free space, neither the optical perception of the distance and solidity of the visual object nor that of the overall perspectival structure of the visual space is possible. For it is the free space in which the objects appear perspectively *arranged* and *structured*, and which in this way gives the visual space its fullness.

The objects in the visual space are depicted on the retina, although this image may be considered completely inadequate in terms of the actual perception of the size, depth and position of the objects and the perspectivity of the entire visual space. However, the free space, in contrast to the material objects, does not leave any kind of image on the retina. The complete absence of free space as an image on the retina seems extremely puzzling to us. If the *input* of the free space or the free-spatial expansion is completely absent as an image on the retina, how does the direct visual perception of free or intermediate space, which forms an important part of the immediate visual field, develop?

An important basis for this puzzle is the following: in contrast to the visual objects, the free space is not a *material object* that can be imaged on the retina by the light rays. The visual free space is an air space on earth and a void in outer space that the light rays penetrate. Thus, the phenomenon of light fills the free space. The perspectival extension of the visual free space is established by the orthogonal structure of the visual light space, or, more precisely, the visual light pyramid. The free space as light space is neither depicted on the retina nor viewed, but is *perceived* in the direct vision, which constitutes the aporia of free-spatial perception. As discussed above, such a *phenomenal aporia*,[30] which can hardly be overcome by the subjective

---

[29] Berkeley (2009, pp. 7–9).

[30] When we stand on the shore of a sea or on a mountain and look at the horizon, we experience how vast the visual space is, stretching to the horizon and the sky. Indeed the visual perception of

thinking alone, seems to *dictate* a certain insight to the subject. This insight would be nothing other than the extra-bodily extension of the sense of sight through the medium of light rays, which *in themselves* are not seen, but which give fullness to the visually perceived free space. In this extra-bodily extension of vision, the subject *feels* the free space directly as a pure, extended nothingness, which is however filled by invisible and orthogonally and perspectively structured rays of light.

The visual perception of free space, which *in itself* is a mere extended nothing, is the direct *visual sensation* of emptiness, which is, however, subject to an orthogonal *gradation.* That is, the *intensity* of the visual perception of space diminishes in proportion to the distance from the eye (or in proportion to the depth of the visual space), which supports the above-discussed thesis of the extra-bodily extension of vision through the orthogonally and perspectively structured light and visual rays. It is our direct visual experience that, in our immediate vicinity, both the free space and the objects are *precisely* perceived in approximately the correct size, distance, extension and solidity. The visually *precise* perception of the free-spatial extension, distance and the solidity of the objects, etc., weakens in direct proportion to the distance from the eye point. However, the retinal image, in which the free space is not depicted at all and the near and far objects are depicted more or less undifferentiated as flat phenomena, cannot contain any input for such a gradation of visual spatial perception. We could denote the properties of our visual space that *enable us* to precisely perceive the size, solidity, distance, etc., of an object as its *intensity.* This intensity of the visual space, in which seeing is an optical touching—that is, a synthesis of the optical and haptic senses—is reduced in direct proportion to the distance from the eye.[31]

This reduction in the intensity of the visual space in distance or depths, as a consequence of which the precision of the perception of visual depths, the free-spatial extension and the solidity of the objects seen becomes less and less, can also be perceived in the virtuality of a perspectival image—a painting or a photograph. However, such a perception arises from our familiarization with the real perspectivity of vision itself. We have already discussed how our direct vision is to be regarded as an optical touching in terms of its real extra-bodily extension. The *haptic extension* of the sense of sight into the depths of the visual space obviously has its limits; the size, distance, solidity of objects, the extension of free space and the perspectivity (or perspectival deformation) of vision are perceived less and less precisely and diminutively in the depths of the visual space compared to our immediate surroundings. This gradual reduction of the *intensity of the visual field* in direct proportion to its depth seems to be caused primarily by the fact that the extra-bodily extension of the sense of sight loses strength and intensity in the depths.[32] Although the phenomenon of

---

the free space, which is not imaged on the retina, *prima facie* constitutes a subjective-perceptual aporia. But this aporia is primarily based on the existence of free space as light space and the equally objective fact that this free space does not leave a retinal image. Therefore, the aporia of the visual perception of free space can basically be described as a phenomenal aporia.

[31] On the intensity of the visual space and its gradation, see Thaliath (2005, p. 221 ff.).

[32] Ibid., pp. 235–236.

light or a possible electromagnetic *nature* of visual rays does not lose its strength at immeasurable distances, the *nexus of the subjective sense of sight* with this physical phenomena could hardly maintain its strength and intensity regardless of the depth of the visual space.

The gradual reduction of the intensity of the visual space in the depths should reach a limit—or a limiting plane—beyond which the sense of sight cannot extend. At this border level, where the intensity of the spatiality or the three-dimensionality of the visual space almost completely disappears, no solidity or depth differences of the objects can be perceived, and the visual objects appear flat and extremely diminutive. Our immediate visual space clearly represents such a limit or border level of vision. At the celestial level, which is obviously the limit of our vision, the visual objects appear flat and very small; neither their huge three-dimensional extension nor the spatial depth differences can be visually perceived. To a lesser extent, objects on the horizon also appear flat and very small to us, and their differences in depth elude our vision. The sky-level or celestial plane as a flat boundary plane of the visual sense can be directly seen or experienced in a clear night sky—with the appearances of the moon, the planets and the stars (but also with the distant clouds that are colourfully exposed in the northern lights).

3.2.8 The Moon Illusion

The moon illusion is a well-known and centuries-old aporia in the theory of visual perception of size, for which no satisfactory explanation appears to have been proffered. The moon appears considerably larger near the horizon than it does higher in the sky at the zenith. Between these two positions of the moon, we experience a gradual reduction in the size of the rising moon. However, this visual illusion applies not only to the moon, but also to the sun as well as to the constellations of stars and planets in the sky.

> The apparent horizon enlargement of the moon is usually called the *moon illusion*, a term which became popular in the twentieth century. It has also sometimes been called the *horizon illusion*, but this is clearly less appropriate. A similar illusion can be observed for the sun, and it is normally called the *sun illusion*. This illusion is less well known to most people than the moon illusion, perhaps because the sun is usually too bright to observe with naked eye. However, the sun illusion was more frequently discussed than the moon illusion in the early literature. Although two different celestial bodies are involved, there seem to be no fundamental difference between the two illusions: they are therefore generally considered to be two examples of the same phenomenon. A third example is the apparent enlargement of the constellations and of the distances between the stars near the horizon. In the words of the astronomer Paul Stroobant (1884, p. 719): 'The same phenomenon exists for the constellations; thus the Great Bear and Orion, close to the horizon, appear enormous.' This form of the illusion is probably the least observed of the three, since fewer people give

serious attention to the night sky. The term *celestial illusion* has been proposed to cover all three examples.[33]

The above discussion makes it clear that the aporia of visual size perception as represented in the moon illusion is a general phenomenon in all celestial appearances. This celestial illusion is primarily an illusion of visual size perception, but its *occurrence* in the sky also suggests another related illusion, i.e., that of visual distance perception of celestial phenomena. Although the horizon appears to be far away from an observer on the earth as compared to the zenith, the moon, the sun or the star constellations appear considerably larger near the horizon, which contradicts our visual experience on earth, where distant objects appear small. According to this illusion of visual distance perception, the sky appears to us to be *flattened-dome-shaped,* like a flat-domed cupola, although it should appear spheroidally domed. The flattened-dome-shaped appearance of the firmament, which causes the above-mentioned celestial illusions to *transpire,* forms the basis for a well-known and thoroughly topical theory of the moon illusion, namely the thesis of the flattened firmament proposed by Abu Ali Al-Hassan Ibn al-Haytham (c. 1039 AD).

The fact that the observation of the sun or moon illusion and the attempts to explain them have a very long history indicates that it is one of the earliest *aporias* to be passed down in the history of philosophy and science. The earliest known references to the sun illusion found in various oriental and occidental cultures date to the seventh century BC in Egypt, the fourth century BC in Greece, and the first and second centuries AD in China.[34] Aristotle tried to explain the phenomenon of the sun illusion with his thesis of refraction; however, the philosopher also advocated the theory of visual rays, which was generally accepted in his time in ancient Greece.

---

33 Ross, Helen E. & Plug, Cornelis (2002): *The Mystery of the Moon Illusion. Exploring Size Perception*, Oxford University Press, Oxford, p. 1.

34 See Ross, Helen E. & Plug, Cornelis (1989): *Historical Review, The Moon Illusion*, ed. Maurice Hershenson, New Jersey, pp. 5–7. In their introductory remarks on the history of the moon illusion, Ross and Plug outline various observations of this optical phenomenon—not the moon illusion, but the analogous sun illusion—in the Orient and in Greece. Worth mentioning are the reports from China, which correspond with the ancient observations of celestial illusions, in particular those of Ptolemy: "Another early prescientific reference to the illusion occurs in Chinese manuscript of possibly the 1st or second century AD. The author is claimed to be Lieh Tzu, an imaginary philosopher who supposedly lived during the fourth century BC. The document relates an apocryphal story about the philosopher Confucius, who found two boys arguing about the distance of the sun. One said that the sun was much larger at sunrise than at noon and must therefore be closer to us in the morning. The other answered that the sun is much hotter at noon and must therefore be closer to us in the middle of the day. Confucius was unable to solve the problem (Giles, pp. 68–69). This story indicates that the sun illusion was well known in China at the time of writing, and that the sun's angular size was thought to be measurably enlarged, due to its being closer. [...] The astronomical chapters of the official history of the Chin dynasty (256–420 AD), written in the seventh century under the supervision of emperor T'ai Tsung, contain a discussion by the scholar Ko Hung (283–343 AD) of the theory that the sun cannot be seen at night because it moves a great distance away from the earth. Ko Hung then argues (Ho Peng Yoke, 1966): 'If it is said that the sun can no longer be seen because it has gone a very great distance from us, then when the sun sets, on its way towards the north, its size should diminish. But, on the contrary the sun becomes larger at sunset. This is far from being proof that the sun is moving further away when it sets'" ibid., pp. 6 and 8.

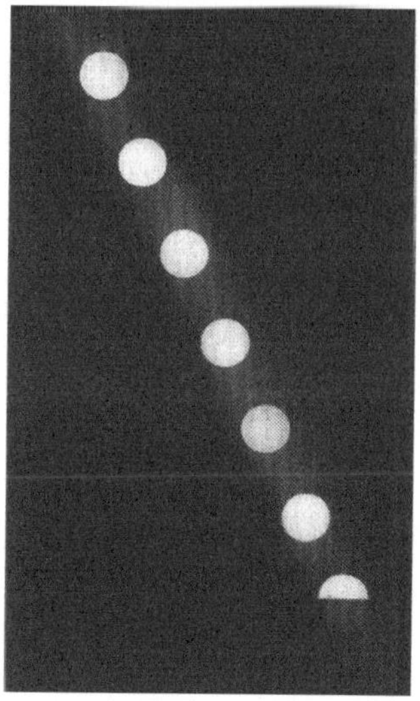

Fig. 3.13 Photographic evidence of the moon illusion. Ross, Helen E. & Plug, Cornelis (2002). *The Mystery of the Moon Illusion. Exploring Size Perception.* Oxford University Press. Oxford. p. 60

Aristotle's theory of refraction was echoed in the later ancient theories of the sun illusion in Greece and Rome—particularly in the work of Claudius Ptolemy. These ancient theories were revived in the Arab cultural sphere, particularly in the works of Farghani (ninth century), Jabir ibn Aflah (twelfth century) and Ibn al-Haytham (eleventh century).[35] Here it is important to mention the early medieval astronomer Ibn al-Haytham (known as Alhazen), whose *theory of the flattened dome* in his major work *Optics* had a great influence on the early modern investigations of the lunar illusion. The Renaissance and the early modern period fell back on the problem of the lunar illusion, which had remained unresolved since ancient times and beyond the Arabic and European Middle Ages. Several attempts to solve this aporia were undertaken by many philosophers and scientists of optics in the early modern period (Descartes, Kepler, Galileo, Gassendi, Hobbes, Berkeley, Molyneux, Euler, among others), and also by artists or artist-philosophers such as Leonardo da Vinci.[36] In the nineteenth century, the moon illusion was interpreted by well-known philosophers, mathematicians and scientists such as Carl Friedrich Gauss, Hermann von Helmholtz, Wilhelm Wundt and others.

The aporia of the moon illusion is based on the already discussed variation of the visual perception of the size of the moon in the sky, irrespective of the constancy of the input, namely the size of the retinal image in the eye of the observer. The moon appearing larger near the horizon and its apparent smaller size higher in the sky give rise to a retinal image whose size or diameter remains constant. This implies that the visual angle formed by the large appearance of the moon near the horizon and its smaller appearance high in the sky with the observer's eye point remains unchanged.

[35] Ibid., p. 9.

[36] Ibid., pp. 10–15.

With the development of photography towards the end of the nineteenth century, this problem could be easily demonstrated.

> The horizon enlargement of the moon is so compelling that its illusory nature has to be pointed out repeatedly. The constant image size of the sun and moon can be demonstrated by photographs in addition to quoting astronomical measurements. Photographs were mentioned in this connection by Haenel in 1909. These illustrative devices sometimes inspire more confidence than the most accurate measurements. For example, following Luckiesh (1921), Hamilton (1964, p. 490) declared that ›since the invention of the camera, it has been proven that the moon is no larger at the horizon‹. This type of evidence was later published by Solhkhah and Orbach (1969) and by Walker (1978a). It consists of a print from a single negative, on which the rising moon or sun has been repeatedly photographed while keeping the camera in the same position, thus recording a series of (equally sized!) images over a period of time (Fig. 3.13, added by the author).[37]

The gradual reduction in the size of the moon on its way from the horizon to the zenith can be seen with the naked eye, but cannot be proven photographically. The constancy of the size of the retinal image of the moon (or the sun) indicates the constancy of the *input* in the process of vision, which, however, paradoxically produces a variation of the actual apparent size of the moon in the sky as output. This raises the question of whether the variation of the visual size perception in the moon illusion develops internally, i.e., solely in the domain of physiological optics, or whether it originates externally in the actual visual space. The changing size of the moon that is immediately apparent when viewed directly, while the size of its retinal image remains

[37] Ross and Plug (2002, pp. 59–60). The authors also explain the photographic illusion in the familiar (photographic) representations of the moon illusion: "Unfortunately the camera can also lie. Many postcards are on display showing grossly enlarged photographic images of the sun or moon. These pictures are usually taken with a telephoto lens, and show a distant horizon with the moon subtending the same angular size (half a degree) as a building on the skyline. The reason the pictures are misleading is that they show no foreground, and therefore appear to represent a close shot of near buildings but with the moon subtending the same angular size as a near building. (…) Alternatively, if a standard lens is used and the foreground is shown to scale, a separate image of the enlarged moon can be superimposed on the horizon scene (Fig. 3.14, added by the author). ibid., pp. 60–61.

Fig. 3.14 Photographic misrepresentation of the moon illusion, Ibid., p. 61

constant, tends to indicate the latter. The moon illusion illustrates like no other perception of virtuality the *dichotomy* between the internal psycho-physiological and the external geometrical-optical *process of seeing*.

In the course of history, various attempts have been made to resolve this dichotomy in the process of vision, on which the aporia of the moon illusion is based. These attempts were either limited to the internal psycho-physiological part of the process of vision or extended to its external geometrical-optical part. Ptolemy's theory of refraction, handed down through centuries, is a prime example of an attempt to explain the outer geometrical-optical part of the vision by means of a dioptric light phenomenon. In modern times, however, in keeping with the *Cartesian* subjectivism, there has been a tendency towards physiological and psychological "cues" that could produce the moon illusion as an optical *virtuality*. The theory of *intermediate objects*, which was introduced in the modern era and is still being discussed today, explains the psycho-physiological origin of the moon illusion as an optical virtuality. According to this thesis, the visual objects *on a horizontal plane* between the eye and the moon near the horizon suggest the optical illusion of the enlarged appearance of moon. The well-known thesis of perspectival illusion also belongs to this category of explanation. When the moon is higher in the sky, these "cues" that supposedly suggest the enlarged visual appearance of the moon or the sun near the horizon are missing. However, even when the intermediate objects as "cues" are methodically negated, the moon illusion at the horizon will remain. In his treatise *An Essay towards a New Theory of Vision*, Berkeley describes an experimental test of this kind:

> Many attempts have been made by learned men to account for this appearance. Gassendus, Descartes, Hobbes, and several others have employed their thoughts on that subject; but how fruitless and unsatisfactory their endeavours have been is sufficiently shewn in *The Philosophical Transactions*, where you may see their several opinions at large set forth and confuted, not without some surprise at the gross blunders that ingenious men have been forced into by endeavouring to reconcile this appearance with the ordinary Principles of optics. Since the writing of which there hath been published in the *Transactions* another paper relating to the same affair by the celebrated Dr. Wallis, wherein he attempts to account for that phenomenon which, though it seems not to contain anything new or different from what had been said before by others, I shall nevertheless consider in this place.
>
> His opinion, in short, is this; we judge not of the magnitude of an object by the visual angle alone, but by the visual angle in conjunction with the distance. Hence, though the angle remain the same, or even become less, yet if withal the distance seem to have been increased, the object shall appear greater. Now one way whereby we estimate the distance of anything is by the number and extent of the intermediate objects: when therefore the moon is seen in the horizon, the variety of fields, houses, etc., together with the large prospect of the wide extended land or sea that lies between the eye and the utmost limb of the horizon, suggest unto the mind the idea of greater distance, and consequently magnify the appearance. And this according to Dr. Wallis, is the true account of the extraordinary largeness attributed by the mind to the horizontal moon at a time when the angle subtended by its diameter is not one jot greater than it used to be.
>
> With reference to this opinion, not to repeat what hath been already said concerning distance, I shall only observe, *first*, that if the prospect of interjacent objects be that which suggests the idea of farther distance, and this idea of farther distance be the cause that brings into the mind the idea of greater magnitude, it should hence follow that if one looked at the horizontal moon from behind a wall, it would appear no bigger than ordinary. For in that case the wall

> interposing cuts of all that prospect of sea and land, etc. which might otherwise increase the apparent distance, and thereby the apparent magnitude of the moon. Nor will it suffice to say the memory doth even then suggest all that extent of land, etc., which lies within the horizon; which suggestion occasions a sudden judgment of sense that the moon is farther off and larger than usual.[38]

Even without such an experimental attempt, the inadequacy of the Wallian thesis, against which Berkeley polemicizes, can be understood. In a city or in the countryside, the moon appears near the horizon *behind* several objects such as houses, streets, fields, trees, mountains, etc. But when we are standing on the beach and watch the moon rise or the sun set on the horizon, there is no *intermediate object* (except for a flat horizontal plane of water) that is supposed to suggest the moon or sun illusion (which we can see directly and clearly in this case).

It is important to note here that Wallis's concept of a *natural* nexus between the visual perception of size and the *actual* distance of the visual object from the eye clearly implies the extra-bodily extension of vision. "We judge not of the magnitude of an object by the visual angle alone, but by the visual angle in conjunction with the distance. Hence, though the angle remain the same, or even become less, yet if withal the distance seem to have been increased, the object shall appear greater." In ordinary vision, the size of a distant object always appears smaller to us. Wallis's idea corresponds to the appearance of the moon *at a visual threshold*. When Wallis observed that the moon appeared larger as it approached the horizon (which is the actual distance and as such does not need to be suggested by intermediate objects) even though the angle of vision remained unchanged or, in the case of the moon illusion near the horizon, became smaller, he tacitly ascertained the *actual* geometrical-optical basis of the moon illusion—that is, its extra-bodily mode of existence. This is because both the angle of vision and the distance of the visual object are directly related to the orthogonal *real* light rays, i.e., to the light rays reflected by the object and converging on the eye point, and thus to the real extra-bodily extension of the sense of sight.

In his comment, Wallis does not refer to the actual distance of the moon when it is near the horizon, but to the suggestion of the distance of the visual object through the intermediate objects. But in reality, the moon near the horizon is at a greater distance from the observer. However, the depth of distant objects such as celestial bodies, which is immediately *seen,* does not depend on their actual distances from the observer on the earth. Nevertheless, the suggestion of size through the distance of the moon, as presented by Wallis, seems to presuppose a certain *preformation* or *pre-existence* of the visual boundary plane discussed earlier, which most likely comes about through our visual familiarization with the nature that surrounds us. If, following Wallis, the factum of the distance of celestial phenomena is added to the process of vision—in particular to the visual perception of size—a certain form of the real boundary level of the visual space emerges, which for us—the observers—becomes recognizable *simultaneously* through the distance and through the size of the celestial phenomena. The well-known theory on the moon illusion, which refers

---

[38] Berkeley (2009, p. 34–35).

directly to such a limiting level of vision, comes from the Arab astronomer Ibn al-Haytham (Alhazen) in the eleventh century: the theory of the flattened firmament.

> One of the most popular theories of the moon illusion relates the horizon enlargement of the celestial bodies to the perceived form of the sky. Many people see the sky as a flattened, dome-shaped surface. If the celestial bodies are seen as moving along this surface, then the sun and moon should appear closer at higher angles of elevation than towards the horizon. But a close object of a given angular size is smaller than a distant object of the same angular size. Therefore the sun and moon should appear smaller overhead than on the horizon.[39]

Following this consideration, it can be seen that the celestial plane, on which the moon and the sun move, creating the illusion of size variation, is a visual boundary plane; however, it is not a real but a virtual plane. Likewise, the distant appearances of the moon and other celestial bodies are virtual appearances in which they appear *separate from the real objects*. The immense distance of these celestial bodies from the earth cannot be visually experienced, because direct vision only extends to a certain limiting plane of the sky; according to the variation of this distance, the visual boundary plane appears to us as a flat firmament. The extremely small appearance of the heavenly bodies (in comparison to their considerable sizes) at the visual boundary level can be attributed to their very small image on the retina and to the relatively small distance that the visual field can reach. The question now is whether our visual *limiting plane* in the form of a flattened firmament, which appears to arise from the geometrical-optical principles and to which the moon and sun illusion can be traced, is a real or a virtual phenomenon. Alhazen's original theory of the flattened firmament does not refer to a *real one*, but to the visual perception of the distance of the moon near the horizon as suggested by the intermediate objects:

> The flattened dome theory was first clearly formulated by the Arab scientist Ibn al-Haytham (*c.* 1039) in his *Optics*, though it was repeated in very similar terms by the thirteenth-century writers Bacon, Pecham, and Witelo. The part on the moon illusion consists of several pages of repetitive discourse. The passage can be summarized as follows. The size of an object is judged by combining its visual angle with its known distance. The distance can be properly judged only if the observer can see an uninterrupted sequence of intervening bodies between himself and the object. When looking up at the sky there are no intervening objects – we perceive only an expanse of colour. This appears to us like a blue surface. Because we are familiar with plane surfaces such as walls, the sky is seen as a flat plane extending in all directions. Such a plane must appear closer directly overhead than towards the horizon. The celestial bodies appear to move along this plane and are therefore judged to be further away when on the horizon than when overhead; consequently they are seen as larger on the horizon.[40]

Alhazen and the later proponents of the *theory of the flattened firmament* assume that the intermediate objects suggest to the observer, when looking horizontally, that the moon near the horizon is farther away and therefore appears larger. However,

---

[39] Ross & Plug (2002, p. 99).

[40] Ibid., pp. 99–100. Also see Ross and Plug (1989, p. 17). "Ibn al-Haytham, and the medieval authors following him, argued that intervening objects towards the horizon cause the sky appear flattened, which in turn causes the moon and sun to appear further, and therefore larger, on the horizon."

there is an essential difference between these two suggestions. The first suggestion of distance—through the intermediate objects and through the perspectivity of the visual space in which they participate—is rather a pure psychological-optical phenomenon, whereas the second suggestion, in which the perception of distance prompts the subject to perceive the *corresponding* size, is primarily a *real geometrical-optical construction.* In real or direct vision, the distance of the object suggests a reduction in its appearance (like trees on a distant hill or a ship on the horizon). This would mean that our immediate visual space—regardless of whether it is merely constructed by the brain or really extends extra-bodily—is subject to two *objective* principles of optics. If we do not consider it a theoretical inadmissibility to start from the direct visual experience of the moon or sun illusion, it could be concluded that the geometrical-optical structure and laws of the moon illusion—according to the *theory of the flattened firmament*—are based on a *real extra-bodily* extension of the sense of sight. It is the *actual* form or extension of the boundary plane of vision that produces the moon or sun illusion, or the larger appearance of these celestial bodies near the horizon and their gradual diminution in the firmament, when they move towards the zenith.

The moon illusion is based on both the clear difference in size between the moon at the horizon and the moon at the zenith as well as on the gradual reduction in the size of the moon as it *moves* from the horizon to the zenith of the sky—along an approximately elliptical path on an equally elliptical inner plane of the firmament (as shown and immediately apparent in Fig. 3.16). The fact that the entire occurrence of the moon illusion on the celestial plane corresponds to the geometric form of the visibility of this plane—as a flattened dome—and, in addition, to the immediately experienced variations of distance in vision—between the eye of the observer and the appearance of the moon moving from the horizon to the zenith—, speaks for a real or extra-bodily occurrence of the moon illusion as a *geometrical-optical* phenomenon. The observed form of the sky seems to be directly *given;* it is not suggested by the intermediate objects that only occur in horizontal vision. The factum of the intermediate objects can easily be negated here without the visible shape of the sky as a flattened dome or the moon or sun illusion disappearing. Even if we are standing all the time behind a wall that blocks the visibility of all intermediate objects up to the horizon (as Berkeley suggested), and we follow the path of the moon from the horizon to the zenith under a clear sky, it is possible to experience the moon illusion or the gradual reduction in the size of the moon's appearance. Standing in the centre of a vast plain, the evening sky, in which the still and distant clouds are colourfully illuminated by the northern lights, would appear to us distinctly in the form of a flattened dome. Likewise, the clear night sky, *without the visibility of earthly objects between us and the horizon*, appears in the form of a flattened dome, where we *can see* the celestial bodies—the moon, the planets and the stars—spread out on a *sky plane*.[41] The visible form of the sky as a flattened dome—even without considering

[41] When we are standing on the shore of a sea or in the centre of a vast plain—fields, lakes or even playgrounds—we see the evening or night sky clearly as a flattened dome. From the centre of a vast plain or green open space such as *Parker's Piece* in Cambridge, for example, a cloudy evening sky,

the sun or moon illusion on the sky—where we see the celestial phenomena on the horizon far away compared to those high at the zenith, would mean that the extent of the boundary plane of our visual space is already *preformed*, which basically determines the *form* of the visible celestial plane with scattered appearances of the celestial bodies and celestial illusions or even *causes* them.

The development of such an immensely extended visual plane can be traced back to our everyday habits. We usually see *frontally* and *horizontally*. We stand, walk and sit (all of the most common body positions in everyday life) with our bodies held upright and our eyes facing frontally and horizontally the surrounding space and objects *on the ground*. Due to gravity, our postures are most of the time vertical; even when we sit, our upper body is vertical. Most phenomena on earth are vertical due to gravity—trees grow vertically and masons, while constructing a wall, place a layer of bricks vertically on top of another layer of bricks, i.e., aligning them vertically using a *plumb line* in accordance with the law of gravity. That is why all trees, all natural phenomena, as well as man-made structures, stand vertically on the ground and face us, who also stand vertically on the ground and see frontally. Unlike most animals, fishes and birds, humans have a special physique: they stand upright on two feet, their heads held high, their eyes not directed downwards towards the ground, as in most animals, but looking horizontally, in line with the vertical position of the body. In addition, both eyes are positioned frontally in the head. In animals, fishes and birds, the eyes are mostly not frontal, but on the side of the head. The frontal positioning of the eyes in the head enables humans to have stereoscopic vision, which largely perfects the optical perception of space.

As Fig. 3.15 shows, in our lives we mostly encounter vertical phenomena, which, because of the vertical positioning of our body and head when standing or sitting—as well as when moving—and because of our frontally aligned eyes, we always see in elevation, and in fact in perspectival frontality and depth. Due to gravity and the verticality of our body structure, most vertical phenomena on the earth are imaged on the retina by the eye lenses, which are equally vertical and directed frontally. The retinal image is therefore vertical and frontal in the geometrical-optical structure of vision, and the eye axes are mostly horizontal. Humans, who live on the earth and have developed a frontal perspective in the course of their evolution, develop a different perception of space than birds, especially birds of prey and migratory birds, which usually see the objects on the earth not in elevation but in plan view and whose

---

colourfully lit by the northern lights, appears not as a spherically domed sky, but as a flattened or approximately ellipsoidally domed sky, where the clouds on the horizon appear far away compared to the clouds high in the sky. Likewise, a clear night sky (when all objects between the horizon and us are invisible) appears like a flattened dome with the moon, the planets and the stars. If we connect the celestial bodies that appear everywhere with *imaginary* lines, we see a network of celestial phenomena that appear to be scattered on the inner plane of the flattened firmament. However, a clear blue daytime sky also appears to us in the shape of a flattened dome. On the horizon, we see the blue sky far away, whereas the blue boundary level of the sky above our heads appears relatively close. In principle, the blue boundary level of the sky (which is caused by the dispersion of light in the free airspace) should appear to us as a spheroidal bulge at a constant radial distance. These direct empirical observations seem to support the traditional *thesis of a flattened firmament* with regard to celestial illusions *under all circumstances*.

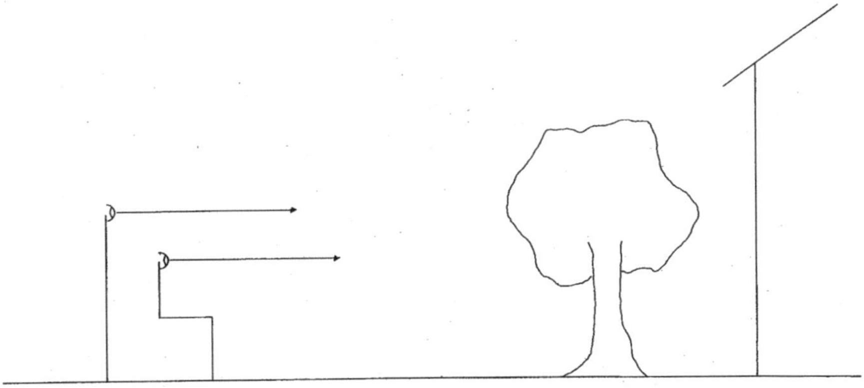

Fig. 3.15 The horizontal-frontal vision[43]

eyes are not directed frontally but laterally. While we have become accustomed to a predominantly horizontal perception of distance, these birds learn to perceive the vertical distance, i.e. the depth of the objects visible in the ground plan on the earth, much more precisely, so that they can orientate themselves correctly in space and time when landing or striking their prey. A person who is accustomed to a frontal-horizontal perception of distance finds it difficult, for example, as in case of a pilot while landing an aircraft, to correctly estimate the vertical distance, in this case the depth of the ground. James J. Gibson explains this particular case that prompted various psychological research programmes, especially during the Second World War, ultimately leading to the ground theory of visual space:

> At the beginning of World War II there was a sudden need to understand the perception of depth and distance as they applied to aviation. The critical task of estimating distance from the ground when a flier is landing an airplane was particularly important. Research was begun and studies of space-perception multiplied, based on what psychologists already knew about it from the great experiments of the 19th century. A list of the *clues* or *cues* for the perception of the distance of an object had resulted from these experiments and this list had gained acceptance. The cues were classified as binocular or monocular according to whether they depended on the use of two eyes or one eye.[42]

The pilot of an aircraft that is slowly landing cannot see the intervening objects on the ground between his eye and the runway horizontally, like the way he would be able to see the horizon on the ground horizontally through all intervening objects. While landing the aircraft, he sees the objects on the ground in an oblique perspective, but he sees them primarily in a plan view; the most important object between the eye and the runway is therefore the free space, which—as a purely extended nothing—cannot, unlike material objects, suggest any *additional* perception of depth. Accordingly,

[42] Gibson (1904, p. 6).

[43] Thaliath (2001, p. 124).

the pilot in the aircraft is hardly able to estimate the oblique vertical distance to the ground—compared to his distant horizontal depth perception on the ground.

All these considerations illustrate that the long habituation to horizontal and frontal vision on earth prompts the human eye to construct a special form of its visual space or visual boundary level, in which horizontal depth perception develops greatly in comparison to the mere vertical and intermediate oblique depth perceptions. The reason for this would be the intensification of the visual space; with the natural habituation to horizontal seeing, a *far-reaching* intensity of the visual space could develop, which produces the greatest depth of the visual boundary plane on a horizontal visual plane. In contrast, the laborious and infrequent or unaccustomed vertical vision seems to be very limited or to lose intensity of depth perception. Consequently, the visual boundary plane cannot achieve greater depth. The *gradation* of visual depth perception from the horizontal to the vertical plane of vision then forms the entire extension and structure of the visual space and the visual boundary plane, most clearly evidenced by the *unchanged* visible form of the sky as a flattened dome. The objects in between and the stereoscopic vision contribute to the visual depth perception, but they at most form the so-called "cues" that do not originally produce the essential factum, namely the visual depth perception, but only modify an already *existing* depth perception.

If we compare the immediate surroundings of our visual space, where all visual objects are perceived in approximately correct size, solidity and distance along with the extension of the free space, with the real visual boundary plane of the sky on which the celestial bodies *appear*, we realize that the *intensity of the visual space* discussed above decreases orthogonally in the depth of the visual space, completely disappears at the visual boundary level of the sky. It is from such a *limit* of the continuous reduction of the intensity of the visual space that the immediate visual boundary level, on which neither the solidity nor the differences in distance of the visual objects (or celestial bodies) can be perceived and beyond which the visual space does not extend, is said to *really* arise. What would be the reality or the real phenomenality of such an orthogonal reduction of the intensity of the visual space and its complete disappearance, from which the immediate visual perception of space and the visual boundary plane arise? This obviously cannot be achieved by real light rays, which can reach immeasurable distances, but rather by the visual rays, which, while seeing, seem to form a structural unity with the light rays. For it is plausible to assume that the *projective* visual rays orthogonally penetrating the visual space—apart from their true mode of existence as a purely subjective or physical-material phenomenon—lose strength and intensity at greater distances and thus construct a visual boundary plane.

The existence of visual rays was a generally accepted basic concept in ancient theories of optics. The emission of visual rays from the eye and their structural unity with the physical light rays, from which the process of vision develops, were presented and discussed by Plato (of course *metaphorically*) in the *Timaeus*.[44] The

[44] Plato (1997): *Timaeus*, trans. Donald J. Zeyl, in: Plato. Complete Works, ed. John M. Cooper, Hackett Publishing Company, Indianapolis, p. 1248 (45 b–e).

pre- and post-Platonic theories of vision by Pythagoras, Aristotle, Euclid, Ptolemy, among others, although they differ in their determination of receptivity (from object to eye) or projection (from eye to object), imply the existence of a medial connection between the eye and the objects of vision. Euclid, Aristotle and Ptolemy attempt to explain the celestial illusion—in particular the solar illusion—in terms of the *production* and *projection* of the visual rays. The basic argument is: the effort or strain of the eye when looking up at the sky *makes it more difficult* for the visual rays to flow freely out of the eye; consequently, the visual rays *cannot reach far* compared to the case where we effortlessly look horizontally. The result of this optical strain of the eye is the smaller and relatively closer appearance of the sun or moon high in the sky, which is obviously subject to the simple geometrical-optical structure of the visual space:

> Ptolemy has been credited also with a completely different explanation of the moon illusion. This appears in a book entitled the *Optics* (Ptolemy, 2nd cent. AD), which exists only as a 12th-century translation by Eugene of Sicily into medieval Latin, from an Arabic translation of what was presumably a Greek original. It is not clear how much of the contents were written by Ptolemy and how much is later additions.
>
> The passage relating to the moon illusion states (Ross & Plug, 1976):
>
> 'For since it is a universal law that, when the visual ray strikes objects of sight in unnatural and unusual conditions, it has a reduced sensation of all their characteristics, similarly it will have a reduced sensation of the distance which it covers. This would seem to be why, amongst objects in the sky which subtend equal angles between the visual rays, those near the point above our head seem smaller, whereas those which are near the horizon are seen differently and under normal conditions. But objects high in the sky seem small because of the unusual conditions and the difficulty of the action. (p. 379).'
>
> There are substantial difficulties in interpreting this passage (Ross & Plug, 1976). In terms of the visual ray theory of vision it might mean the following. When looking upwards, and in other unusual viewing conditions, it is more difficult to push one's visual rays out to the object being observed. This is in line with Aristotle's claim (ca. 322 BC/1962, p. 478) that illness makes it more difficult to push one's visual rays out. Because of this difficulty, all visual sensations, including those of size and distance, are reduced (for a reason that is not stated). An object overhead, when compared to another of the same angular size seen in a horizontal direction, will therefore appear nearer and smaller.[45]

In their commentary on this astonishing observation by Ptolemy, Ross and Plug point out that the Ptolemaic explanation does not provide a precise explanation for the reduced perception of size and distance of the celestial bodies high in the sky ("for a reason that is not stated"). But the reason for the size and distance reduction of celestial phenomena is very clearly implied in the Ptolemaic explanation. In the unnatural way of seeing, or rather in the strenuous act of looking up, the visual rays projected by the eye—due to the physical and ocular exertion—are *unable to reach far compared to normal and effortless looking towards the horizon;* consequently, the visual rays form *geometric-optically* closer and smaller celestial appearances high in the sky. The geometrical-optical form and laws of our *extra-bodily visual space*, on which the Ptolemaic explanation is based, fundamentally correspond to the geometrical-optical representation of the theory of the flattened firmament (Fig. 3.16), where the visual

[45] Ross and Plug (1989, pp. 8–9).

Fig. 3.16 Moon illusion in a flattened dome-shaped sky[46] Ibid., p. 61

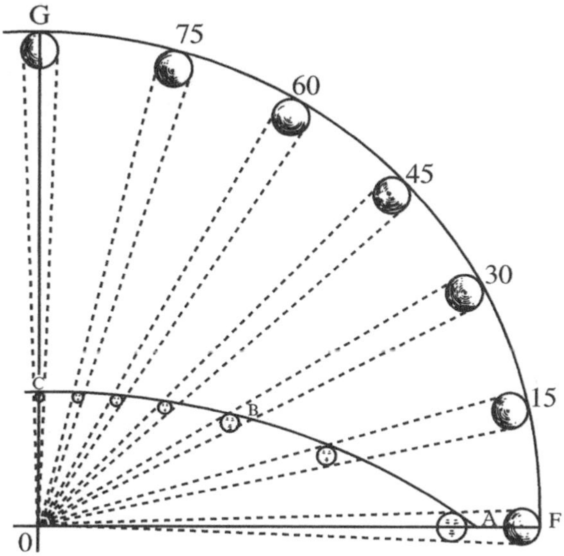

rays, which become shorter and shorter from the horizon to the height of the sky, produce both the gradation of the size reduction of the celestial appearances—from the horizon to the zenith of the sky—and (accordingly) the flattened shape of the sky or the *flattened-dome-shaped* boundary plane of the sky, through their different distances of penetration in the celestial space.

Several experimental visual exercises conducted from the early modern period to the present report the disappearance of the moon illusion when the eye is strained. Many common and popular accounts of the moon illusion point out that the illusion disappears when the moon is seen in an unusual position of the head, i.e., when the body is bend over and the moon viewed from between the legs:

> It has often been noticed that inverting the head makes objects look smaller and reduces the horizon enlargement of the moon. Some authors have put this down to proprioceptive effects, and others to changes in the visual scene when inverted. Sanford (1898, pp. 211–12) commented that colours appear distorted, and that distances may appear either enhanced or reduced. Some authors have commented on the resulting size-distance paradox. For example, Angell (1932) remarked: 'More pronounced is the effect of gazing with head inverted; objects appear decidedly smaller and likewise farther away, the latter probably a »secondary illusion« which connects smaller size with greater distance.' The effect was also reported by Meili (1960), who noted that the perceived enlargement of distant hills was much less when tilting the head or looking through one's legs: he argued that this was because size constancy depended on the presence of a good *gestalt*, or well-structured visual scene, and tilting the head somehow changed this. Coren (1992) made a rather similar point: he argued that an inverted pictorial scene reduces the 'salience' of depth cues, and depth cues rather than proprioceptive factors were the cause of the moon illusion. Explanations involving an inverted visual scene have even crept into modern popular literature, as in a novel by Atkinson (1997). Boring (1943) also commented that the horizon moon was much diminished

[46] Ross and Plug (2002, p. 100).

> when viewed through one's legs, and ascribed this to changes in the angle of regard or the vestibular system. [...] The effect of various bodily postures was first investigated by Gauss (1880). Gauss found that when his body was tilted backwards so that the elevated moon shone in his face, the moon appeared larger than when he looked up by tilting his head and turning his eyes up. Thus movements of the head and eyes seem to be implicated, though Gauss's limited observations cannot be conclusive. Later investigators have repeated Gauss's experiment rather more formally, with several observers. Reimann (1902b, 1904) made numerous estimates of the sizes of the sun and moon from various bodily positions, but neither he nor any of his observers found any effects of posture. However, Zoth (1902), using 50 observers, confirmed Gauss's results.[47]

Another experiment frequently tried in the early modern period was looking at the moon on the horizon through a tube. This visual exercise led to the perception that the lunar illusion disappears on the horizon:

> During the seventeenth century authors began to speculate about the outcome of 'tube' experiments: what would happen to the perceived size of the horizon sun or moon if the surrounding objects were blocked from view? The idea for a tube experiment may have come from a statement about atmospheric effects by the Greek astronomer Cleomedes, around the third century AD: 'It is also said that where it is possible to view the sun from deep wells, its appearance is much larger since it is seen through the humid air of the well.' Riccioli's contribution (1651) is of interest, because he seems to have been one of the first to suggest an experiment. He disbelieved in the importance of intervening objects. He noted:
>
> "If you observe the rising or setting sun in your bedroom or garden in such a way that a hedge or wall or lower edge of the window prevents all view of the intermediate space up to the horizon, and that you see nothing else but the sun, you will however see it as immensely greater than when far from the horizon."
>
> Other authors suggested looking through a tube to remove the effect of intervening or surrounding objects. The contributions of Molyneux (1687) and Goüye (1700) were mentioned in Chapter 7: Molyneux asserted that the illusion remained, Goüye that it disappeared. [...] The outcome of a genuine tube experiment is important for theories of the illusion. If the illusion is eliminated when the horizon moon is viewed through a tube, the enlargement must be due to intervening or surrounding objects – that is, to perceived distance or to relative size, or to luminance or colour contrast. If the illusion remains at full strength, it must be due to one or more other factors. If it occurs, but at reduced strength, it must be due to a combination of factors that includes the visual surrounds. Despite the importance of the experiment, there was no agreement among early authors on its outcome. [...] The effects reported are, of course, generally in line with the authors' main explanation for the illusion. On the whole, those observations that are apparently genuine suggest that the illusion is eliminated or much reduced when the horizon moon is viewed through a reduction tube.[48]

In these visual experiments, the function of the tube through which the moon was viewed on the horizon was to effectively eliminate the intervening objects between the horizon and the observer. But the intervening objects when looking at the moon on the horizon can be eliminated during natural viewing without the moon illusion disappearing—as Berkeley's experiment discussed above shows. Furthermore, the moon illusion proves to be a continuous process, or a gradual reduction in the size of the moon on its *way* from the horizon to the zenith in the sky. If we observe the gradual

---

[47] Ibid., pp. 171–172.

[48] Ibid., pp. 118–119.

reduction in the size of the moon in the sky from a seashore, we see the complete moon illusion on the horizon and its gradual disappearance as the moon rises in the sky, not through any intervening object. Therefore, the disappearance of the moon illusion when looking through a tube seems to be based on a different optical fact. This fact could be the loss of strength and intensity of the visual rays—*analogous to* the cases discussed earlier—in contrast to their *free* emission in a completely open visual space in normal vision. In direct vision, the *free* and immeasurable expansion of the visual space enables the eye to see the visual objects equally free, i.e., with all the strength and intensity of the visual space; the narrowing of the visual space when looking through the tube then causes the loss of this intensity of the visual space and consequently the reduced perception of the size of the moon on the horizon (because the visual rays, reduced in strength and intensity, do not reach far). Likewise, the strain of the eye while looking through the tube, which is analogous to the other types of experimental straining of the eye described above, could lead to the disappearance or correction of the moon illusion.

The normal straining of the eye, which most closely matches the immediate perception of the moon illusion, that is, the gradual reduction in size of the moon on its way from the horizon to the zenith in the sky on a flattened inner plane of the firmament, is due to the effort of looking upwards. The results of many experiments in history show how the moon illusion reappears when this natural type of straining is normalized. In one such natural experiment, people climbed a mountain to look again horizontally at the rising moon (like looking at the moon on the horizon), which appears smaller from the land or sea level. The result was the return of the moon illusion:

> A more convincing field experiment to prove that the proximity of the moon to the visible horizon is the determining factor in the illusion was performed by Lohmann (1920). He first observed the enlarged rising moon while standing with his back to a hill. When the moon had risen some distance, the illusion disappeared. Lohmann thereupon climbed over the top of the hill to view the moon from the other side, above a new and higher horizon. In this position the illusion reasserted itself.[49]

The importance of this experimental evidence lies in the fact that it corresponds to our everyday observation of the sky illusions and—in connection with this—with the *constant* appearance of the sky as a flattened dome. It is our everyday experience that the sky—irrespective of its appearance as a clear daytime blue sky, cloudy twilight sky in the evening or starry night sky—appears much farther along the horizon than the sky high above our heads. The immediate and always visually experienced form of the sky as a flattened dome more or less coincides with our everyday experience of

[49] Ross and Plug (1989, p. 19). In this context, an experimental observation by Karl Friedrich Gauss is worth mentioning: "When I view the elevated full moon from a backwards reclining bodily position, in which the head is in the usual position with respect to the body, so that the moon shines more or less perpendicularly on the face, then I see it much larger; and conversely I see the full moon on the horizon noticeably smaller with the head inclined forwards." ibid., p. 16. See Gauss, Karl F. (1880): *Briefwechsel zwischen Gauss und Bessel* (herausgegeben auf Veranlassung der Königlich Preußischen Akademie der Wissenschaften), Leipzig, pp. 498–499 (trans. Ross and Plug).

the celestial illusions, i.e., the gradual reduction in size of the appearance of the moon or the sun on their way from the horizon to the zenith, and thus seems to confirm the theory of the flattened firmament, passed down in the history, in its *geometrical-optical* principles. Now, the observation of the celestial illusion at the celestial level, i.e. in the sky, points to the reality or the real phenomenality of the extra-bodily extension of the sense of sight—up to the celestial level—which presupposes the equally phenomenal reality of the material mediation *in the visual space* between the eye and the visual boundary level of the sky.

Another feature of the moon illusion is the varying size perception of the moon on the horizon. Many observers report that the magnified appearance of the moon on the horizon is not constant. There are days when the moon appears larger on the horizon than on other days; on some days, an oversized appearance of the moon on the horizon has been observed. Such a change in the size of the moon can hardly be explained in psycho-physiological terms. It is inconsistent to assume that individual observers have different psycho-physiological dispositions of their sensory abilities on different days, so that they perceive the moon on the same horizon in different sizes on different days, even though the ophthalmic input, that is, the size of the retinal image, remains unchanged. It is even more difficult to comprehend the case when several observers on different days have the same experience of the size variation of the moon illusion. The fact that the size variation of the moon illusion on the horizon is *simultaneously* perceived by several observers clearly indicates that this phenomenon does not result from internal or psycho-physiological factors, but from external circumstances. These external circumstances could be atmospheric variations, which push back our familiar boundary plane of the visual space extending to the horizon, resulting in the increased size of the appearance of the moon on the horizon. The enlargement of the moon's appearance on the horizon could *suggest* a virtual proximity of the object of vision, as a result of which the moon near the horizon appears closer to the observer.

The constant shape of the sky as a flattened dome—the basis of Alhazen's thesis that has been handed down in history—seems to be the most important fact here that could unlock the mystery of the moon illusion and the sky illusion in general. For the mode of the moon illusion, that is, the gradual reduction of the size of moon's appearance from the horizon to the height or zenith in the sky, empirically proves this thesis. Now the seemingly preformed expansion of the sky as our visual boundary plane and the immeasurable expansion of our visual space are to be considered *real* or *phenomenally objective* and not just as a psycho-physiological suggestion. The flattening of the appearance of the sky, which should *in principle* appear to us as a spheroidal dome, can be attributed to the variation in the *intensity*[50] of our visual field,

---

[50] In my doctoral thesis, completed in 2003 (title: *Perspektivierung als Modalität der Symbolisierung.* Erwin Panofskys Unternehmung zur Ausweitung und Präzisierung des Symbolisierungsprozesses in der *Philosophie der symbolischen Formen* von Ernst Cassirer), I attempted to demonstrate geometric-optically the intensity variation of our immediate visual space within the framework of a theory of perspectivisation. My investigation is based on Erwin Panofsky's seminal essay *Perspective as Symbolic Form.* I tried to prove that the geometrisation of the *perspectiva naturalis* into the pictorial-architectural *perspectiva artificialis*, as conceived by Panofsky, can basically

which in turn seems to be based on the fact that we are used to seeing our environment frontally and horizontally. The celestial plane seems to provide theoretical evidence for the *actual* extra-bodily extension of the sense of sight, as does the previously discussed visual perception of size—a fact that was recognized in ancient times but was suppressed or even completely rejected in the Cartesian modern age in favour of the apriority of the spatial intuition. The prevailing epistemic internalism of early modernity, which still persists, or rather triumphs in the theory of perception, seems to have established the legitimacy of the intromission theory of vision. This form of perceptual internalism is tacitly regarded as an infallible paradigm that would prevent the validation of any attempt to *objectively* externalize the process of visual or auditory perception—both conceptually as well as experimentally.

3.2.9 The So-Called 'Cues'

In the history of ophthalmology, various "cues" have been pointed out that particularly *suggest* visual depth perception. Ophthalmologists, psychologists and other scientists in ophthalmic optics are still working on these *cues*, which to a large extent form the bodily and extra-bodily factums such as the movement of the eye and of objects, stereoscopy, perspectivity, etc., and *believe* that they have *sufficiently* explained the enigmatic phenomenon of visual depth perception on the basis of these *cues*. As Gibson rightly observed, despite all attempts at the explanation in history, visual depth perception remains an unsolved mystery, an aporia that defies our powers of comprehension. The best-known of all the *cues* that suggest three-dimensionality and the immeasurable depth of visual spatial perception from a simple input such as the *flat* retinal image is stereoscopy or binocular vision. The factum of stereoscopy in visual depth perception is supported above all by the well-known stereoscopic virtuality, where we can, by means of an ocular-optical technique, perceive the solidity

be interpreted as a geometrical-optical perspectivisation in the process of seeing, which results in the intensification of the aesthetic visual space. The basis for this kind of perspectivisation is the geometrical-optical *flattening out* of the originally spheroidal or spheroidally distorted retinal image into a plan-frontal image, which results in the intensification of the visual space through the visual rays that are *similarly intensified* or *straightened*. The process of geometrical-optical perspectivisation, from which our visual space emerges, gives rise to the gradation of the intensity of the visual space. Accordingly, the intensity of the visual space varies or decreases *orthogonally*—towards the depth of the visual space—and frontally, i.e. on the plane perpendicular to the visual axis. This property of geometrical-optical perspectivisation in visual spatial perception contributes directly to the higher intensity of the familiar *horizontal* perception of depth, as discussed above. This is because, according to geometrical-optical perspectivisation, the visual space in the center or along the visual axis is of higher intensity than at the periphery of the visual space. Therefore, we are able to see much deeper through the visual space on a horizontal plane and along the usual *horizontal* position of the visual axis than on a vertical plane. This seems to explain the preformation of the expansion of the sky as our visual boundary plane in the form of a flattened dome. On the process of geometrical-optical perspectivisation in visual spatial perception, see Thaliath (2005, p. 213 ff), and on its geometrical-optical demonstration, ibid., pp. 351–367.

and depth of the objects depicted on a flat image much more thoroughly than in a perspectival image.

These *cues* can indeed perfect the visual perception of depth and solidity, but it would be quite wrong to assume that they *alone* suggest or create the visual perception of depth. This is because they can be systematically negated without the visual perception of depth being lost. It seems that the cues are improvising an *already existing ability* of visual spatial perception, which forms a basic phenomenon and the foundations of which are yet to be discovered. In their research on visual depth perception, scientists tend to refer to these *cues*. However, they are manifold and do not constitute the *sufficient* causes of visual depth perception. Therefore, they do not provide science with the *definitive* solutions, prompting scientists to continue their research in this area with a relatively large selection of *research objects*.

When science searches for the fundamental cause of a scientific aporia—which is the essence of fundamental research—the *cues* must not be considered as conclusive or final research subjects. Favouring one of these *secondary facts* as a research subject ultimately blocks the research process and prevents the discovery of the fundamental principle that would wholly explain the scientific and phenomenal aporia. In research, therefore, the search for a fundamental principle should take precedence over the diversity of *cues*—the secondary facts that appear to be the only key to the scientific puzzle. The process of scientific *investigation* of phenomena in the research should therefore aim at a fundamental explanatory principle before it refers to other secondary facts that build on this principle. If the research process initially focuses on one of the secondary facts, the fundamental explanatory principle—that is, the primary, deeper and original cause—will most likely remain unexplained. In fundamental research, this *necessary disregard* of all secondary facts in order to determine the fundamental and as such irreducible explanatory principle is accomplished by means of the method of systematic negation, which, as is well known, was realized in the history of modern philosophy and science by the Cartesian method of doubt and negation.

The above-mentioned cues do not actually *cause the origin* of the visual depth perception. However, by improvising or even perfecting visual depth perception to a large extent, they can be regarded as secondary causes. These secondary causes must first be systematically negated in order to *isolate* a primary cause as the fundamental explanatory principle of this perceptual aporia. Now we will examine how this necessary *empirical* research process can be put into practice. The *cues* that have been recognized as objects of research and are currently being investigated in the study of visual depth perception are:

1. stereoscopy
2. movement of the eyes, the pupil
3. movement of the objects
4. perspectivity

All these so-called *cues* that scientists employ to explain the phenomenon of visual depth perception can be methodically negated. We see the *still* objects in the visual field spatially or in the free spatial distance and solidity with *a still* eye. We look at objects with one eye through a camera, a telescope or a microscope and perceive them spatially, regardless of whether they are still or moving. Such monocular vision is imperfect compared to the stereoscopic vision, but it is originally a *spatial vision*. This means that the factum of spatial vision does not originate in this *cue*, but in a property of the visual process that precedes it.

Likewise, the perspectivity of the visual space does not prove to be an *absolutely necessary* condition for spatial vision. In our ordinary or everyday visual experience, objects appear to us in a perspectival spatial structure or in a perspectival depth, position, size and extension, but for that reason alone we cannot ascribe to the perspectivity of our seeing the status of the originator of visual spatial perception. Furthermore, perspectivity arises from the orthogonal convergence of light rays, i.e. from a physical dimension of depth, which as such has a perceptual-theoretical priority over perspectivity. Just like the other *cues* of the visual spatial perception mentioned above, the perspectivity of seeing can be methodically negated. For example, consider that I am sitting in my room at midnight and looking out of the window. Outside it is completely dark—except for a tiny lamp glowing somewhere far away. In this situation, I see only a punctual illumination in a certain position and distance through the window and no other objects that could form a perspectival spatial structure and thus give the visual space a perspectival depth. Here I have the direct visual experience that a beam of light—or a very thin bundle of light rays—emanates from a distant *point* in complete darkness and reaches the retina through my eye point. I see the punctual source of the light in the correct position and at approximately correct distance in a very large and completely dark space. To this scenario of everyday visual experience, in which I dispense with the perspectivity of my visual space, I can add the above-discussed negation of the other *cues* of the visual spatial perception. Especially when I dispense with the factum of stereoscopy, my visual depth perception becomes imperfect, but it does not disappear. In the above-described visual experience of depth, where all familiar *cues* of visual spatial perception are dispensed with, I have the immediate impression that my sense of sight *extends* out of my eye *linearly* along a beam of light to the punctiform source of the light in a completely dark space. We have similar visual experiences when we are standing in a vast field on a very dark night and see one or more tiny and punctual light sources far away, or when we see a punctual light source on a far shore from a ship. In such a dark environment, when we see a tiny, stationary point source of light far away with a stationary eye, our visual perception of depth, which now excludes all recognized optical *cues*, can only come about through the direct mediation of light rays. This is because only a point of light appears on the retina in this case, which alone cannot suggest a spatial distance. Under these circumstances, we can see several point light sources that are located at different distances and in different positions in approximately correct positions and distances, although these phenomena do not form a perspectival visual space-structure. All these considerations illustrate that a fundamental principle of the direct extra-bodily extension of vision precedes any so-called bodily and extra-bodily *cue*—on the basis of which scientists attempt to explain the visual depth perception.

3.3 The Extension of the Auditory Sense

In addition to the sense of sight, the auditory sense or the sense of hearing is largely extended outside the body. We hear voices, sounds or noises spatially extended or localized in objects whose mechanical vibrations *originally* cause them. We hear the ticking of a wall clock or the voice of a speaker *localised* in the objects, i.e., in the clock on the wall and in the speaker's mouth, regardless of our position in the room. This shows that the extension or spatial localization of the auditory sense is determined by the objects themselves that are heard and does not depend solely on the listener subjectively and a priori.

Analogies can be drawn between the structures of extra-bodily senses, for example, between the structures of the sense of sight and the sense of hearing. While the light rays or waves *image* the objects in the visual field on the retina, the airwaves *transfer the* vibration of the objects to the eardrum during the process of hearing. Strictly speaking, neither the image on the retina created by the light rays nor the vibration of the eardrum created by the air waves have any input of the free-spatial distance of the (seen and heard) objects. Just like the image on the retina, the vibration of the eardrum is the only *bodily* input in the process of hearing. As in stereoscopic vision, hearing with both the ears perfects the sense of hearing and its extra-bodily, objective localization. An important difference between visual and auditory perceptions is the nature of the mediation of the light rays and that of the airwaves themselves. While light rays or waves are an electromagnetic phenomenon, airwaves, which are produced by objects vibrating in the air, are purely mechanical in nature.

How can the purely mechanical vibration of the eardrum suggest the exact extra-bodily *direction* and *spatial localization* of the auditory sense? The airwaves from different objects reach the ear drum from different directions. Even if we seek to trace the directional perception in the sense of hearing back to this mechanical contact between the airwaves and the eardrum, we tacitly assume an extra-bodily extension of the sense of hearing through the airwaves. But the auditory perception of direction cannot suggest the distance of the heard object in space, i.e., the distance to the source of the voice or sound. Hearing with both ears—just like stereoscopic vision—can indeed perfect the spatiality or objective localization of the sense of hearing, but it cannot create it *originally*. Many scientists claim (and try to prove) that the spatial localization of the sense of hearing is suggested by the tiny time difference between the vibrations of the two eardrums. However, we can negate this factum by appropriately positioning our face in relation to the object we are hearing. It is also possible to hear with only one ear without the spatial localization of the auditory perception disappearing. Just as we can visually perceive the distances of objects with one eye, we are able to perceive the voices, sounds and noises localized in objects with one ear—albeit imperfectly.

In the bodily or haptic or gustatory sensations, their objective localization is perceived in the body itself. Just as with the tactile sensation of cold and warmth, as well as with the pain sensation localized in the foot or the taste sensations localized in the tongue, the mechanical vibration of the eardrum originally forms a bodily phenomenon from which the sense of hearing arises. But we do not hear voices, noises or sounds localized in the body or in the ears, but in extra-bodily objects. If the vibration of the eardrums *without* the airwaves that cause the vibration is the only *bodily* input in the hearing process, I see no reason why I cannot locate someone's voice in my ear and not at the source, i.e., his mouth. Here, as in the case of the sense of sight discussed above, we are confronted with an aporia of the sense of hearing: how does the extra-bodily localization of the auditory sense in objects come about when the only bodily input in the process of hearing, namely the vibration of the eardrums, is only a mechanical phenomenon that does not *involve* any spatial distance of the heard object?

If we assume that the extra-bodily spatiality of the sense of hearing, like that of the sense of sight, arises solely through neuronal processes in the brain, we are once again confronted with the problem of the complete ontological emergence of sensibility. There are causal neuronal states and processes in the brain that are *necessary* for the extra-bodily localization of auditory sensations and that vary according to the different distances of the localization of the auditory sensations. But they can only correspond to the actual auditory experience, namely the extra-bodily spatial extension of the sense of hearing, or only form a causal *correlation* or participation. That is to say, they alone cannot guarantee the factum of the extra-bodily-objective localization of the sense of hearing. In other words, neuronal causality is necessary for the reality of the extra-bodily extension of the sense of hearing, but it is not sufficient. This is because the brain processes involved in hearing are based on the mechanical vibration of the eardrums by the airwaves, where no input is given that could suggest the spatial localization of the sense of hearing.[51] The spatial structure of the extra-bodily sense of hearing has a significantly different reality or phenomenality in contrast to the physiological neuronal processes on which the sense of hearing is based. As in the case of the sense of sight discussed above, the domain of causality seems to extend beyond the *ontological boundary* of neuronal causality into the extra-bodily *reality*. This means that the primary cause of the extra-bodily extension and localization of the sense of hearing is not to be found within the domain of the neuronal processes that serve as the physiological basis for the sense of hearing, but

[51] As with vision, scientists would also work on the many cues in the hearing process—in the hope that they could trace the spatiality of hearing back to these cues. These cues would be the tiny time difference between the vibrations of the eardrums (i.e. the fact of hearing through both ears), the different intensities of the vibrations (due to the air waves coming from differently localised sources), etc., as discussed above. But all these cues can be methodically negated without us ceasing to hear spatially. Regardless of the intensity of the vibration, the sources of sounds, voices, noises, etc. in the near vicinity are perceived much more precisely than those at a greater distance. A very loud explosion in the distance can cause a more intense vibration on the eardrums compared to the soft ticking of the wall clock in the immediate vicinity, but the spatial localisation of the source of the explosion is perceived as less precise than the localisation of the ticking in the clock. This leads to the conclusion that the *actual real* distances of the objects heard are crucial for the localisation of the sense of hearing and not just the differences in the intensity of the vibrations in the air and consequently on the eardrums.

in the reality of the extra-bodily sense of hearing itself, or in a physical phenomenon in the real extra-bodily airspace.

The voices and noises from the objects that we hear in our environment are connected to our ears by airwaves. However, unlike light waves, airwaves are not an electromagnetic but a purely mechanical phenomenon. Therefore, as described above, we can hardly assume that the air waves in the hearing process form an extra-bodily extension of the electrical phenomenon on which all neuronal processes in the brain are based, and thus ensure the extra-bodily localization of the sense of hearing—analogous to the bodily localization of the senses of touch or taste. This supports the previously discussed speculation of a second possibility, namely that the air waves falling on the eardrums give rise to certain electromagnetic waves, which are then emitted by the body and merged with the air waves, simultaneously reaching the auditory objects and thereby—like a sensor—suggesting the extra-bodily localization of the auditory objects to the perceiving subject. The aporia of the extra-bodily extension of the sense of hearing—just like that of the sense of sight—presupposes the necessary existence of a phenomenal mediation of this kind, in which sensibility extends outside the body. The reality of our sense of hearing implies in this way that hearing an extra-bodily and spatially distant object is at the same time an *auditory touching*, from which the spatiality of our sense of hearing arises.

We have already discussed one of the essential characteristics of vision, namely the reduction of the intensity of the visual space in depth, within the context of the spatiality of vision. The same trait also applies to the sense of hearing, or rather to the extra-bodily extension of the auditory space. The localization of the sense of hearing is much more precise for close auditory objects (voices, noises or sounds) than for distant ones. For very distant auditory objects, the sense of hearing loses a great deal of precision and distance perception. Compared to vision, the sense of hearing varies, i.e., the intensity and precision of the sense of hearing reduce much more rapidly, so that we have only a vague perception of the distance of distant auditory objects, even if they generate stronger or more intense vibrations in our eardrums—as in the case of an explosion. This extra-bodily gradation of the intensity of the auditory space can hardly be attributed to the bodily factum of the vibration of the eardrums, which is the only input in the process of hearing. Rather, this property of the sense of hearing—as in the case of the sense of sight—seems to be based on an objective phenomenon by which the auditory sense extends extra-bodily.

Even with nearby objects, the localization of the sense of hearing becomes less precise if our eyes are not directed at these—visible or hidden—objects. If someone is standing behind me and talking, I can still perceive the spatial localization of the voice, or rather, its spatial distance. When I turn around and see the face, my auditory perception of the localization of the voice becomes much more precise. That is, in this case, I perceive the voice as being localized much more clearly in the mouth of the speaker. Similarly, I hear the sound of a fountain at a spatial distance, even if I do not see it (if it is behind me or hidden in a bush). However, if I see the fountain directly, I hear its sounds precisely and localize them exactly at the water level, where the water splashed upwards falls back. Through these and similar everyday experiences, I have the immediate impression that I perceive the extra-bodily localization of my sense of hearing in objects not only auditorily, but in a composite sensibility, i.e., auditorily-optically, and with greater precision. This factum of sensory perception is usually taken for granted in everyday life, where we assume that the sense of sight contributes to the perception of the objective localization of the sense of hearing and makes it more precise. However, the distance seen differs significantly from the distance heard, so that a purely *subjective* suggestion of the auditory perception of distance through the sense of sight is hardly possible. Rather, the *actual* and extra-bodily extension of the sense of sight and that of hearing *on the same object* of these sensory perceptions seem to *intersect* and thereby enable the subject to perceive the objective localization of the sense of hearing more precisely.[52] If this extension of extra-bodily sensibility arises—as previously speculated—through the visual rays and auditory waves emitted by the body, or rather by the eyes and ears, they *intersect* in an object that is both heard and seen. The increased precision of the objective localization of the sense of hearing would thus develop through this intersection, which is basically of a *mechanical* nature. In this respect, such a crossing of the visual and auditory senses in the perceptual object forms a simultaneous optical and auditory *touching* of this object and points to a *real* extra-bodily extension of sensibility.

This speculation can be confirmed by another aspect of hearing. To perceive the spatial localization of the auditory sense more precisely, we do not necessarily have to see the object of hearing during its mechanical vibration; it is enough to *direct* our gaze at the source of the sound, voice or noise. In a radio with a metal frame, we do not see the hidden loudspeaker and the vibration of its membrane, but we are able to locate all the songs, speeches, etc., that are transmitted from the device where the loudspeaker is hidden. Consider the case: we go for a walk and then hear the sounds of the church bells behind us and perceive their approximate location. We turn around and then direct our gaze to the place where the church is hidden behind many trees. Here we see neither the church nor the bells hidden in the steeple, but by directing our gaze in this way, we are able to perceive the source or localization of the sounds much more precisely.[53] Here we see how the sense of hearing is analogous to the sense of sight in many of its extra-bodily properties. Through the two forms of sensory perception, our sensibility extends outside the body. Such an extension or

---

[52] See Thaliath (2008, p. 39).

[53] Even if we assume that vision suggests the precision of hearing only subjectively and a priori—and not on a phenomenal basis—we can hardly explain the primary sensory phenomenon, namely the auditory perception of the localisation of an immediate auditory object, in terms of a transcendental subject and the purely a priori status of space—as a form of sensibility. In everyday life, we experience the spatial localisation of the sense of hearing largely in invisible auditory objects, which also means that our auditory perception of the auditory object and its spatial localisation *occur first*, before we become aware of it or its existence. The *primary* auditory perception of the localisation of the auditory objects or sources is therefore determined more by the object itself. Just as in the case of the direct reference to the object in the visual size perception, as discussed above, here the subject seems to have an immediate extra-bodily and, as such, purely objective reference to the distance and spatial localisation of the auditory object before consciousness—or before an act of consciousness.

embodiment of senses seems to arise more objectively or through a bodily and extra-bodily phenomenality. It is less likely that a transcendental subject merely a priori creates such an extension or embodiment of sensibility. That is, it is the *factum of the object* and not that of the subject that primarily determines the bodily and extra-bodily extension of senses. If the object, which exists independently of the subject, is included in the realm of sensibility, it would fundamentally change the traditional status of sensibility (Sinnlichkeit) in the context of (Kantian and post-Kantian) *transcendental philosophy*. This is not a matter of the object taking precedence over the subject, but rather of a certain equalization of these fundamental factums, which together—in the context of a possible epistemology—make up our aesthetic, i.e., sensory experience.

References

Aristotle (1995). *Metaphysics*. In: *The complete works of Aristotle*. Jonathan Barnes (Ed.). Vol. II. Princeton University Press. Princeton.

Berkeley, George (2009). *An Essay towards a New Theory on Vision*. Philosophical Works. Michael R. Ayers (Ed.). London.

Braunstein, Myron L. (1976). *Depth Perception Through Motion*. New York.

Breidbach, Angela (2017). *Übertragene Körper. Diskurse des Schattens im Werk von Hans-Peter Feldmann, W. G. Sebald und William Kentridge*. Verlag Wilhelm Fink. Paderborn.

Gauss, Karl F. (1880). *Briefwechsel zwischen Gauss und Bessel* (published at the initiative of the Königlich Preussischen Akademie der Wissenschaften). Leipzig.

Gibson, James J. (1904). *The Perception of the Visual World*. Greenwood Press. Connecticut.

Jacobi, Klaus (2008). *Kann die Erste Philosophie wissenschaftlich betrieben werden? Untersuchungen zum Aporienbuch der aristotelischen »Metaphysik«*. In: *Metaphysisches Fragen. Colloquium über die Grundform des Philosophierens*. Paulus Engelhardt and Claudius Strube (Eds.). Böhlau Verlag. Collegium Hermeneuticum. Bd. 12. Köln – Weimar – Wien.

Leisegang, Gertrud (1954). *Descartes Dioptrik*. Meisenheim am Glan.

Morgan, Michael J. (1977). *Molyneux's Question. Vision, Touch and the Philosophy of Perception*. Cambridge University Press. Cambridge.

Panofsky, Erwin (1991). *Perspective as Symbolic Form*. Christopher S. Wood (Trans.). Zone Books. New York.

Panofsky, Erwin (1998). *Die Perspektive als symbolische Form*. in: *Deutschsprachige Aufsätze II*. Karen Michels und Martin Warnke (Eds.). Akademie Verlag. Berlin.

Plato (1997). *Timaeus*. Donald J. Zeyl (Trans.). in: *Plato. Complete Works*. John M. Cooper (Ed.). Hackett Publishing Company. Indianapolis.

Ross, Helen E. & Plug, Cornelis (2002). *The Mystery of the Moon Illusion. Exploring Size Perception*. Oxford University Press. Oxford.

Ross, Helen E. & Plug, Cornelis (1989). *Historical Review, The Moon Illusion*. Maurice Hershenson (Ed.). New Jersey.

Thaliath, Babu (2001). *Der Prozeß der Geometrisch-Optischen Perspektivierung in der Visuellen Raumwahrnehmung*. FreiDok (Freiburger Dokumentenserver of the University of Freiburg). Freiburg i. Br.

Thaliath, Babu (2005). *Perspektivierung als Modalität der Symbolisierung. Erwin Panofskys Unternehmung zur Ausweitung und Präzisierung des Symbolisierungsprozesses in der Philosophie der Symbolischen Formen von Ernst Cassirer* (Dissertation). Verlag Königshausen & Neumann. Würzburg.

Thaliath, Babu (2008). *Die Räumlichkeit der Empfindung*. FreiDok (Freiburger Dokumentenserver of the University of Freiburg). Freiburg i. Br.

Thaliath, Babu (2009). *The ontological causation.* Journal of Dharma 33 (2008). pp. 33–56, Bangalore.

Thaliath, Babu (2016). *Wissenschaft und Kontext in der frühen Neuzeit*. Verlag Karl Alber. Freiburg i. Br.

Chapter 4
From Subject to *Object*

4.1 Prioritization of Theoretical-Philosophical Foundations: Subject and Object

The "subject" and the "object" form the two elementary domains of philosophy. The *science* of philosophy should refer equally to the subject, which perceives sensibly and cognizes conceptually, and to the object, which is perceived and cognized. In principle, these two elementary domains should be given equal or *equal consideration* when philosophizing. However, the traditions of philosophy in different cultures demonstrate the alternating prioritization of these domains in their history. Initial philosophical speculations tended to focus on the object; the pre-Socratic philosophies were principally concerned with the basic substance from which the great diversity of the phenomenal world arose. In other words, philosophy began historically with a fundamental ontological question: What is the essential, unified and objective factum to which the diversity of phenomena can be *ontologically* reduced? From this fundamental question, the first philosophical speculations arose as natural philosophies, as it happened in Greek philosophy. The Presocratics were natural philosophers, for whom the general focus of their philosophical speculations was the object or the diversity of natural phenomena. While the pre-Socratic philosophies had the above-mentioned fundamental ontological question as a common starting point, they speculated differently about the elementary and unified basic substances of phenomenal reality, such as water (Thales), air (Anaximenes), atoms (Democritus) or the four elements: earth, water, air and fire (Empedocles).[1]

[1] To a certain extent, the beginning of such philosophical speculations can also be traced in the Indian culture. In the oldest philosophical speculations, as presented in the *Upanishads*, *Vedas*, *Samhitas*, *Brahmanas* and *Aranyakas*, the main question that arose was that of the *objective* existence of man and nature (of which he is a part). The fundamental philosophical questions in the *Upanishads* clearly relate to an objective existence, that is, to an absolutely fundamental ontological mode of existence of human beings and nature. The speculation about the five basic elements, namely the

B. Thaliath, *The Embodiment of Senses*, https://doi.org/10.1007/978-981-96-8468-7_4

The historical continuity of the pre-Socratic natural philosophies was clearly interrupted in the Socratic system. Socrates marked a decisive turning point in the history of Greek philosophy, which began above all with his fundamental question about human knowledge. The Socratic turn in Greek philosophy was thus an *epistemological turn* and as such it marked a clear historical transition from the pre-Socratic philosophical prioritization of nature to the investigation of the human mind. The knowing or cognizing subject became the main object of philosophizing, which, according to Socrates, is primarily a midwifery in which the innate and hidden *true knowledge* in human mind is produced or discovered by a *dialectical* procedure. This epistemological turn in the Socratic system had a profound impact on all post-Socratic philosophies, especially on the philosophy of Plato, the most famous disciple of Socrates. Plato is known to have tried a synthesis of the Socratic system and its dialectic methodology with some of the prominent pre-Socratic philosophies, namely those of Pythagoras, Parmenides, Heraclitus, Democritus and others.[2]

Through this synthesis, Plato is able to reformulate the question of being, that is, the question of objective existence, in the context of his *theory of ideas*. Thus, Plato seems to be driven in his ontology, on the one hand, by a clear metaphysical inclination (based on his idea of archetypes in the realm of ideas) and, on the other hand, by a prevailing epistemological methodology, which he apparently adopted from Socrates. In Plato, the *basis* for the hierarchization of being—from unreal shadows of existence to ideas—is obviously its *cognisability*, as it is most clearly expressed in the Allegory of the Cave. The prisoners in the cave mistake the shadows of artificial objects for real objects as they, under the circumstances, are *not able to clearly and distinctly* see or perceive them. The liberation of a prisoner and his *progressive journey* from being visually fixed on the shadows of artificial objects to seeing natural objects on the surface of the earth and finally to the viewing or cognizing of the ideas in heaven characteristically form an *epistemological path of liberation* in Plato's Allegory of the Cave—a path in which an ignorant person progresses from deceptive to true knowledge, that is, to the knowledge of what truly exists. Ideas constitute the truly existing being as—and only when—they are cognized as such. The Platonic theory of ideas unmistakably reveals the inherited Socratic epistemological basis.

In the history of European philosophy—from the Middle Ages to Cartesian modernity—it is difficult to find a transition comparable to the one in the Greek tradition. But the Cartesian philosophy, or rather the Cartesian method of doubt, and consequently the foundation of modern philosophy established an epistemological turn—namely, a turn towards philosophical subjectivism. This transition or turn is characterized not

---

pancabhutas (earth, water, fire, air and ether), from which the entire diversity of nature arose, comes astonishingly close to the speculations of the pre-Socratics discussed above.

[2] Because of this strategic synthesis, Plato was regarded by Nietzsche, who held the Presocratics in high esteem as compared to Socrates and the post-Socratic Greek philosophy, as a 'mixed character' (Mischcharakter), or rather as a philosopher who developed his philosophy from a synthesis of Presocratic and Socratic philosophies. See Nietzsche, Friedrich (2001): *The Philosophy in the Tragic Age of the Greeks*, trans. Alexander Ross, Cambridge University Press, Cambridge, pp. 1–2.

only by the rise of the *philosophy of mind*, but also by the early modern *mechanical philosophy*, of which Descartes is also considered a founder. Descartes' *Les Principes de la Philosophie* initiated the clear separation between the philosophy of mind and the philosophy of nature, which established itself in the early modern age as *mechanical philosophy*. The modern natural sciences, in particular classical mechanics, optics, physics and chemistry, emerged from the early modern mechanical philosophy. In *Les Principes de la Philosophie*, Descartes attempts to establish the science of mechanics philosophically. However, his methodology suggests a *thinking into* the object that is separate from the subject, which later became the basic procedure of experimental science in the context of the early modern mechanical philosophy (carried out primarily in the investigations of Robert Hooke).

According to this, the historical transition from the medieval *philosophia naturalis* to the early modern *mechanical philosophy* was characterized by a clear tendency towards subjectivization or subjective apriorization of phenomenal reality. In her major work *Zwei Untersuchungen zur nachscholastischen Philosophie*, Anneliese Maier discusses a striking example: In the scholastic tradition, the colours of objects were considered a subjective perception, but the object appearing in colour was not excluded from the *reality* of the colour.[3] It was only with Descartes and the Cartesian method of systematic doubt and negation that a complete separation of the factum of the object from the reality of this and other sensory qualia became almost a convention in the modern epistemology. The subjective a priori determination of the sensory qualities was a major concern of the Cartesian method of doubt and separation, by which Descartes sought to establish the complete existential autonomy of the subject from the phenomena, perceived by the senses and cognized by the mind. According to Descartes, the apriority of the sensory qualities sufficiently proves the complete modal and ontological separation of the mind from the body. Consequently, all subjective sensory qualities can be separated from the body. What remains in the body—as its purely objective property—is then, for Descartes, a pure extension, that is, the *res extensa*. The Cartesian method of systematically doubting all sensory qualia and attributes of the object is therefore strategically directed towards isolating the subject from its connectedness to the phenomenal world and thereby towards the complete autonomization of the subject as the *primary mode of our existence*, to which the phenomenal world appears to be *subordinated*. This rather epistemological strategy is clearly evident in many examples of the method of systematic doubt.

In *mediations* Descartes attempts to methodically separate the bodily sensory qualities from the body and thereby ascribe them to the sensibly perceiving mind alone.[4] This step leads programmatically to the propaedeutic principle: "ego cogito,

[3] Maier (1968, p. 18).

[4] In *Meditations*, the Cartesian method of doubt involves the negation and separation of the sensory qualities from the body and extra-bodily objects. At the beginning of this seminal work, Descartes discusses the methodical doubting of bodily sensations (such as pain); the final outcome of the method of doubt is the fundamental propaedeutic principle, 'ego cogito, ergo sum', which refers to the absolute ontological separation of the mind from the body. The further examples also refer to the extra-bodily objects from which the sensory qualities are separated. The best known of these is the wax analogy in the second meditation. See Descartes (2009, pp. 33–35).

ergo sum". Pain is felt in a part of the body—hand, stomach or foot—but it is not this part of the body that feels the pain. Through the nervous system, all parts of my body are connected to my brain. Accordingly, all physical sensations are processed in the brain and perceived as physically localized by means of the nervous system, which is spread throughout the body. Nevertheless, I can hardly determine that it is my brain that feels physically. Descartes uses his method of doubt to prove the status of the "I", the subject, as the author of all sensations. The gradual separation of the body's involvement in sensibility leads to the complete isolation of the mind from the body:

> Es bleiben noch die Wahrnehmungen, die Affekte und die Gefühle, die man zwar auch klar erfassen kann, wenn man sich genau vorsieht und gerade nur das über sie aussagt, was in unserem Vorstellen darüber enthalten ist, und dessen wir uns unmittelbar bewußt sind. Indes ist es sehr schwer, dies innezuhalten, wenigstens in Bezug auf unsere Empfindungen, weil wir alle von Kindheit auf angenommen haben, daß alle die Dinge, die wir empfinden, eine Existenz außerhalb unseres Bewußtseins hätten, und daß sie den Empfindungen oder Ideen, die wir von ihnen bei Gelegenheit ihrer Wahrnehmung hatten, ganz ähnlich seien. Wenn wir z. B. eine Farbe sahen, meinten wir eine außer uns befindliche und eine der Idee dieser von uns wahrgenommenen Farbe ganz ähnliche Sache zu sehen, und infolge der Gewohnheit, so zu urteilen, glaubten wir diese Sache so klar und deutlich zu sehen, daß sie uns für gewiß und zweifellos galt. Ganz ebenso verhält es sich mit allem übrigen Wahrgenommenen, auch mit der Lust und dem Schmerze. Denn wenn auch diese letzten nicht außerhalb unsrer selbst verlegt werden, so werden sie doch nicht in den Geist oder in unsere Vorstellung allein gesetzt, sondern in die Hand oder in den Fuß oder in einen anderen Teil unseres Körpers. Es ist aber durchaus nicht sicherer, daß der z. B. in dem Fuße gefühlte Schmerz etwas außerhalb unseres Geistes ist und in dem Fuße sich befindet, als daß das in der Sonne gesehene Licht auch in der Sonne sich befindet; vielmehr sind beides Vorurteile aus unserer Kindheit.[5]

The proposition *ego cogito, ergo sum* (I think, therefore I am) is the final result of this and similar methods of doubt. When Descartes emphasizes the primacy of the

[5] Descartes, René (1965): *Die Prinzipien der Philosophie*, übersetzt und erläutert von Artur Buchenau, Felix Meiner Verlag, Hamburg, pp. 24–25.
See also Descartes, René (2002): *The Principles of Philosophy*, trans. John Veitch, Blackmask Online, p. 24.
"There remain our sensations, affections, and appetites, of which we may also have a clear knowledge, if we take care to comprehend in the judgments we form of them only that which is precisely contained in our perception of them, and of which we are immediately conscious. There is, however, great difficulty in observing this, at least in respect of sensations; because we have all, without exception, from our youth judged that all the things we perceived by our senses had an existence beyond our thought, and that they were entirely similar to the sensations, that is, perceptions, we had of them. Thus when, for example, we saw a certain colour, we thought we saw something occupying a place out of us, and which was entirely similar to that idea of colour we were then conscious of; and from the habit of judging in this way, we seemed to see this so clearly and distinctly that we esteemed it (i.e., the externality of the colour) certain and indubitable. The same prejudice has place in all our other sensations, even in those of titillation and pain. For though we are not in the habit of believing that there exist out of us objects that resemble titillation and pain, we do not nevertheless consider these sensations as in the mind alone, or in our perception, but as in the hand, or foot, or some other part of our body. There is no reason, however, to constrain us to believe that the pain, for example, which we feel, as it were, in the foot is something out of the mind existing in the foot, or that the light which we see, as it were, in the sun exists in the sun as it is in us. Both these beliefs are prejudices of our early years, as will clearly appear in the sequel."

mode of existence of the "I"—the thinking subject—and posits that it can also exist without a body,[6] we clearly see how the Cartesian method of doubt formed an appropriate propaedeutic to a philosophy of the subject, which significantly influenced and shaped modern philosophies. The human mind became the focus of philosophizing. Cartesian rationalism was immediately met with a critical response in British empiricism, represented by Locke, Hobbes, Berkeley, Hume and others. However, these post-Cartesian philosophers recognized the human mind as the main subject of philosophical inquiry. Hence, almost all the major works of the leading post-Cartesian philosophers arose out of an investigation of the human mind and the cognitive apparatus—in the form of a series of treatises on the human understanding by Locke, Berkeley, Hume and others—including Kant's *Critique of Pure Reason*. To this day, modernity is seen as being dominated by the philosophies of the subject or of the mind.

We have previously discussed how the polemic of Princess Elisabeth of Bohemia against the Cartesian idea of an immaterial and inextended soul resonated with Descartes, and how he, in his reply, admitted the soul's connection to the body in acts of volition and sensory perception.[7] Descartes' confession has historical significance. It signalled the historical transition into a philosophical tendency of modern age, namely the prioritization or even favouring of the subject over the material object to which the body belongs. The connection or union between body and soul, emphasized by Princess Elisabeth and admitted by Descartes, clearly refers to the domain of senses, especially the bodily sensibility. In the Cartesian system, the different forms of sensibility were basically defined as modes of thinking and thus subordinated to a *thinking* subject.[8] This is because the ontological difference between the mind and the body becomes much more apparent in thought—that is, in the domain of a purely logical subject—than in bodily and extra-bodily sensory perceptions. *Linguistic-conceptual* thinking obviously proves to be a completely different mode of being from that of the material body and brain.

Cartesian subjectivism had a decisive influence on the early modern natural philosophy; it seemed to tacitly prioritize those fields of science that were predominantly based on the factum of the subject or subjective apriorism, i.e., sciences in which the subjective notion is directly involved and thus clearly takes precedence over phenomena. In this respect, it was no coincidence that the historical progress of the natural sciences in the early modern period came about in the course of mathematical sciences and developed from *mechanical philosophy*. The original sciences in the early modern period were classical mechanics and geometrical optics—two important mathematical sciences. Both these sciences have objects of study that can be easily mathematized, such as the free space, the static and dynamic structures of bodies, *linear* light rays (which are subject to mathematical-optical or -dioptric phenomena such as reflection and refraction), vectorial forces, etc. These

---

[6] Descartes (2009, p. 85).

[7] See Ebbersmeyer (2015, p. 29).

[8] Descartes, René (1972): *Meditationen*, trans. and ed. Artur Buchenau, Felix Meiner Verlag, Hamburg, p. 145.

mathematical-a priori sciences, although they relate to natural phenomena, were obviously based on the primacy of the domain of the subject over that of the object, especially in the mathematical methods of their axiomatization. The apodicticity of their principles was thus attributed to their a priori nature. Even when *mechanical philosophy* began to change from its original mathematical basis to a material one, it could hardly free itself from the subjective apriorism. The relatively late rise of chemistry as a material science was originally prompted by aprioristic speculations about the basic phenomena such as atoms and corpuscles (by Locke, Hobbes, Gassendi, Boyle, and others).

4.2 The *Factum of the Object* in Sensibility

How can a fair balance be achieved in philosophy between the factum of the subject and that of the object? Cartesian rationalism differentiated philosophy as philosophy of the subject from the natural sciences, which developed a different claim to truth in relation to natural phenomena. Early modern *mechanical philosophy*, which was to establish the desired balance between philosophy and the natural sciences and thus between their objects—that is, between the mind and the matter—has never been able to do so without a strategic prioritization of one of these two domains. Initially conceived in mathematical terms, *mechanical philosophy* evolved into the material sciences, which, in the course of their development, became increasingly estranged from philosophy—its methods, its claims to truth and its legitimation.

Natural phenomena should be preserved as an object of natural science and natural philosophy in their unity; they can hardly be subordinated to pure subjective perspectives. Especially in the case of the human body as a natural phenomenon, in which the factum of the object seems to be intertwined with that of the subject, a balanced or justified treatment of these factums discussed above is of particular importance. Descartes' propaedeutic method of doubt clearly shows a highly subjective approach or perspective; the object of investigation, e.g., the physically or bodily localized sensation of pain, is here considered rather from the subject's perspective and in favour of its autonomization (in relation to the body). In the context of the natural sciences, or more precisely in the context of physiology, this approach can be reversed to a certain extent by first taking into account all the bodily or neuronal processes that take place in the brain and throughout the nervous system, and then adding the factum of the subject that ultimately feels the sensations. Establishing a balance between the subject and the objective body in relation to sensory perceptions in this way has the consequence that the bodily localization and simultaneity of sensory perceptions can be explained better and more appropriately. We have already explained how the essential features of bodily sensibility, such as simultaneity and bodily localization, are determined rather objectively—by the body or rather by the nervous system spread all over the body. When I move my foot, my bodily sensation of pain, or rather its localization in my foot, also moves. This means that the static and dynamic states of the physical localization of my sensation of pain are primarily determined by my foot and are not merely generated in the brain. Descartes attempts to confirm his idea of the purely subjective origin of the sensation of pain, including its physical localization, by means of the example of the phantom pain—the virtual pain sensation that awakens an equally virtual feeling of being physically localized.[9] The spatial localization of phantom pain is generated by the brain alone; that is, it is a sensation of pain in which the body or the hand or the legs are not involved. But phantom pain as sensible virtuality does not exclude the primal sensible reality of the bodily sensation of pain, in which the objective body is directly involved. For it is the original *bodily* familiarization with the bodily localized sensation of pain that later—after the body part has been amputated—merely suggests the phantom pain.

If we temporarily dispense with the philosophical reduction of the bodily sensations, as undertaken by Descartes and in the post-Cartesian tradition, and consider only the scientific basis of these phenomena, we see that it is the full bodily extension of the nervous system that ensures the bodily localization of all bodily sensations and their simultaneity. As discussed earlier, the electrical phenomenon in the neuronal system gives our haptic, gustatory and olfactory sensations and all forms of pain and pleasure bodily extension and localization, as well as simultaneity. If we now add the factum of the subject to this purely scientific basis of bodily sensibility, or ascribe all sensory content to the subject and its domain, we see that the spatiality and temporal simultaneity of the *subjective* sensations are *objectively* determined by the body, or rather by the nervous system spread throughout the body, and the electrical phenomenon on which it is based. This means that the necessary addition of the factum of the subject to the natural scientific mechanism of sensibility does not result in a one-sided or merely subjective creation of sensibility and the apriority of its spatiality and temporality, but inevitably in a bodily basis or in a *real* spatial embodiment of sensibility. In other words: the unity of subjective sensation with its *objective* or bodily extension and simultaneity constitutes the reality of our bodily sensibility.

This unity is by nature a synthetic nexus, that is, a link between the domain of the subject and that of the body, which have the spatiality and temporality as a common basis. That is, the bodily extension of sensibility results from a synthetic nexus between the spatial and temporal extension of the body and that of a purely aesthetic or sensibly perceiving subject. The two modes of being, the mind or subject and the body, are not unified *ontologically* but are linked synthetically.[10] In the synthetic nexus between the subject and the body, which *houses* the individual subject, these absolute ontologically different modes of being achieve a perceptual equipollence. In short, the reality of our bodily sensibility is indeed subjective, but it is *equally* and *truly* extended in the body. Accordingly, the body forms a synthetic nexus between the two modes of embodiment, namely the embodiment of a purely physiological-material phenomenon and the embodiment of subjective sensibility.

---

[9] Descartes (2009, p. 84).

[10] I have examined the synthetic nexus between the purely aesthetic subject and the body—the object of sensory perception in general—in a previous essay—titled: *Domänen des Subjekts* (Domains of Mind). See Thaliath (2006, p. 13 ff.).

The embodiment of senses, which we clearly recognize in all our physical sensory perceptions, now extends to the sphere of extra-bodily sensibility, that is, to the actual space of seeing and hearing. Just as the physical body is necessarily involved in the physical extension and localization of bodily sensations, the extension of the sense of sight and hearing should be based on an *objective phenomenality* that extends outside the body.[11] In the second chapter, we discussed this necessary analogy between bodily and extra-bodily sensibility; it presupposes a phenomenon that is *analogous* to the neuronal process in the nervous system, or to the extremely rapid transmission of electrical signals, as a certain *extension of the nervous system outside the body*. Here, the spatiality of the extra-bodily sensory objects, or rather their optically perceivable position, size, solidity, distance, etc., as well as their audibly perceivable localization, are not determined subjectively and a priori, but are rather given objectively.

This analogy leads to the aporia of extra-bodily sensibility. How does the spatiality of the sense of sight and hearing arise outside the body, that is, without the *direct involvement* of the physical body and its nervous system? If the extra-bodily localization of the coloured appearance of an object, as well as of a voice or a sound, can be seen as analogous to the bodily localization of the sensation of pain or warmth, these extra-bodily sensory perceptions should be supported by an analogous phenomenality—like that of the nervous system—as discussed earlier. But this precondition forms a phenomenal aporia that challenges us to find a speculative solution and, possibly, its objective experimental proof. Such a precondition legitimizes the primacy of philosophical speculation over scientific discovery, in which philosophy, as a *logical* necessity, calls on science to invent a phenomenon *that is definitely present but is yet to be discovered*. This philosophically grounded apodicticity becomes thus the motivation for scientific discovery.

Of all the problems discussed so far, only the case of visual size perception is sufficient to prove the aporia of extra-bodily sensibility. This is because the *problem of reference* is most evident in the visual size perception. We cannot rely on any cues to solve this problem—such as stereoscopy, which can approximately suggest the optical perception of the three-dimensional depth and the solidity of objects seen; moreover, this problem is confirmed once again by the optical phenomenon of *object size constancy*, which Condillac uses to polemicize against the traditional Lockean idea of *unconscious inference* in the optical size perception. The phenomenon of the *object size constancy* also refutes the assumption that the extremely small retinal image in the visual size perception is merely magnified in the *internal*, i.e., cerebral

[11] The *reality* of this extra-bodily extension of sensibility, manifested in the senses of sight and hearing, is based at the same time on the purely mental origin of the secondary sensory qualities, such as colour or sound, and their extra-bodily extension into a spatiotemporal structure of the visual and auditory space. The spatiotemporal structure of the visual and auditory space clearly has the status of primary quality and existence independent of the perceiving subject, and thus the existential autonomy from the subject. Such a *phenomenality* of the extra-bodily sensibility can be explained as follows with the help of an example: While we perceive the objects in colour when we see them directly, these *really existing* objects are actually painted or coloured by our eyes in an equally real visual space. The phenomenality of this visual perception is based, in principle, on a *real* nexus between purely mental qualia, namely colours, and the real objects in the visual space, which exist independently of the subject.

process of vision by a certain factor. The size of the individual objects, as well as their motion—the speed and distance of the movement—appears very diminutive on the retina. But when we actually see, we perceive the movements of objects in our immediate vicinity in approximately correct speeds and spatial distances that the objects cover or the spatial structures they create. It is important to mention here that the extra-bodily extension of the sense of vision, which can provide the immediate reference for the optical perception of static and dynamic quantities, offers the simplest and the most accurate explanation for the visual perception of all movements (in which neither the extremely diminutive image nor its geometrical-optical inversion and lateral reversal on the retina becomes a problem).

To resolve this aporia, we have previously assumed that there could be a *mediation* of the electromagnetic phenomenon between the body and the surrounding space and the objects, which could explain the extra-bodily extension of the sense of sight and hearing. With regard to the necessity that this aporia of the extra-bodily extension of sensibility presupposes, we consider it possible that light, as an electromagnetic phenomenon, could have a function in the emergence of this mediation between the eye and its visual space that goes beyond the retinal image. However, such an assumption is not possible in the case of hearing, because airwaves are not an electromagnetic phenomenon, unlike light rays. Therefore, the assumption of the emission of electromagnetic waves from the body, which form a structural unity with the light rays converging on the eye point in the case of vision and with the airwaves in the case of hearing, seems more plausible. The speculative idea of the structural unification of the projective visual rays and the receptive light rays follows the ancient Platonic thesis of vision; in the case of hearing, this analogy is extended. The basis of this philosophical speculation, which seems apodictic to us, is, on the one hand, the necessity that the aporia of extra-bodily sensibility presupposes, and, on the other hand, the possibility that the electrical phenomenon in the nervous system, which clearly facilitates the bodily localization of the senses of touch, taste and smell, of pain, etc., can extend to the actual visual and auditory space. This is because the electrical phenomenon has the property of being emitted beyond the solid objects in the free space in the form of waves and forming a *field.* Since the electrical phenomenon underlies bodily sensibility spread over the bodily nervous system, especially its bodily extension and localization as well as simultaneity, it would be entirely plausible to assume that the same phenomenon makes it possible for subjective sensibility *in an analogous way* to extend out of the body while seeing and hearing.

So far, this possibility remains a mere speculation, as it has not been experimentally proven.[12] But it is a necessity in relation to the aporia of extra-bodily sensibility. We

[12] The existence of the electromagnetic field, which extends from the body into the surrounding space, has already been experimentally proven. The psychiatrist Colin A. Ross has demonstrated the emission of brain waves from the open eyes in several experiments. As short waves, they are able to reach great distances simultaneously. However, Dr. Ross strangely uses this experimental evidence to investigate the electrophysiological basis of the so-called 'Evil Eye Belief' or an everyday experience of the 'sense of being stared at'. See Ross, Colin A. (2010): *Hypothesis: The Electrophysiological Basis of Evil Eye Belief*, in *Anthropology of Consciousness*, Vol. 21, Issue

are not clear about the true nature or phenomenality of the necessary extra-bodily mediation in seeing and hearing (which establishes the necessary reference to real objects), but we can safely assume *that there must be something extra-bodily* that—in mediation—ensures the spatiality and temporality of extra-bodily sensibility. The fact that an extra-bodily phenomenal mediation has not been discovered in the sense of sight and hearing is also due to the fact that the science of ophthalmology and that of auditory perception are *unwilling to believe* in it. This is because these sciences, within the framework of the Cartesian-modern theory of perception that still prevails, are too fixated on the a priori nature of sensory perception, in which neither the body nor the extra-bodily objects can *directly* participate. In other words, science, due to the handed-down and well-established conviction of the purely subjective origin of sensibility, is hardly inclined to seek an extra-bodily phenomenal mediation as the basis of the extra-bodily extension of sensibility.

But philosophy cannot ignore a clear aporia, which is ultimately a phenomenal aporia. For, as Edmund Husserl emphasized, it is a *strict science*. As such, it tends to search again and again throughout history for the first principles of phenomenal reality, thereby calling into question the axiomatic foundations of the sciences that have long been taken for granted. Accordingly, the leitmotif of philosophy, as already established by Plato, would be to dismantle the doxastic in science or in its explanation of phenomena and to re-establish the episteme. The first step in dismantling the doxastic foundations would then be a plausible philosophical speculation. In reality, the above-discussed extra-bodily extension of the senses of sight and hearing through a *real or phenomenal mediation* would solve all the traditional and barely mastered

---

1, pp. 47–57; quoted from the abstract: "The sense of being stared at is the basis of evil eye beliefs, which are regarded as superstitions because the emission of any form of energy from the human eye has been rejected by Western science. However, brainwaves in the 1–40 Hertz, 1–10 microvolt range emitted through the eye can be detected using a high-impedance electrode housed inside electromagnetically insulated goggles. This signal, which the author calls 'human ocular extramission,' is physiologically active and has distinct electrophysiological properties from simultaneous brain-wave recordings over the forehead." Also see Ross, Colin A. (2011): *Traditional Beliefs and Electromagnetic Fields*, Revista de Antropologia Iberoamericana, Vol. 6,3, Madrid, pp. 274–275: "It is an established scientific fact, then, that EM emissions from the brain and heart travel out into space and can interact with objects in the outside world, such as electrodes. These emissions contain physiologically meaningful information that can be used to make diagnoses and plan treatment in cardiology and neurology. One might think that brain waves and other human EM signals do not travel very far in external space due to the inverse square law, according to which the intensity of an EM signal drops off with the square of the distance. However, the inverse square law does not apply to very low frequency signals in the 1–60 Hertz range, which is the frequency level at which brain waves are measured (Barr, Jones, & Rodger, 2000). At these frequencies there is negligible drop-off in the signal at 1.000 kilometers. Brain waves emitted through the eye can be detected by high-impedance electrodes that make no contact with the body. How far away this signal can be detected depends on the sensitivity of the electrodes and the necessary software development. Human ocular extramission has greater amplitude or intensity than the general field emitted through the skull because it does not have to pass through the skull. For this reason, and perhaps due to conscious focusing, the geometry of the skull, and the presence of the optic nerve terminal in the retina, the signal emerging through the eyes has greater amplitude and a beam-like distribution in space."

problems of extra-bodily sensibility at once. These are, to summarize the discussions so far, the following:

1. Visual perception of size
2. Visual perception of depth, perception of the solidity of objects
3. Visual perception of free or intermediate space
4. Visual perception of position (perception of the upright position of phenomena)
5. Perspective structure of the visual field
6. Visual virtuality—of reflection and refraction—and its geometrical-optical laws
7. Visual perception of movement
8. The extra-bodily objective localization of the sense of hearing

If scientists assume that the only inputs for the processes of vision and auditory perception are the retinal image when seeing and the vibration of the eardrums when hearing, and seek to trace the reality of the sense of sight and hearing *solely* back to the neuronal processing of these inputs in the brain and to a few associated physical and non-physical cues, all these problems remain unsolved—or insufficiently investigated. If a basic assumption, even if it is temporarily an unproven speculation, solves all the identified aporias of a phenomenon—and does so with the utmost simplicity and clarity—that assumption is capable of scientifically proving itself to be a true fact. But what prevents scientists from proving this speculative assumption and thereby legitimizing it scientifically is in all likelihood a factum of belief, because they tend to regard this speculation as hard to believe. However, scientists implicitly rely on a conviction that the sense of sight and hearing and all their properties are caused solely by neuronal processes. On the one hand, purely neuronal causation is insufficient due to its (previously discussed) ontological difference from the *actual reality* of seeing and hearing, and on the other hand, it can hardly explain the aporias discussed above.

When bodily and extra-bodily sensibility has a unified basis in terms of spatial extension, it forms an immeasurably large embodiment. We are immediately aware of the embodiment of our bodily sensibility in our body itself; in seeing and hearing, our sensibility extends into great depths, heights and widths. The embodiment of our sensibility thus extends in a vast visual and auditory space, filling the immeasurably extended optical and auditory free space and including the sizeable and most extensive objects—mountains, lakes, forests, valleys, all the way to the sky and the horizon. Since this embodiment of my sensibility consists in the expansion of my aesthetic subjectivity, it can be said that "I" *expand* through my sense of sight and hearing in my entire visual and auditory space. To us, the heirs of Cartesian modernity, such a notion seems miraculous and rather incredible. But it would be one of the secrets of nature and the cosmos that we could not believe in for thousands of years.

In the embodiment of sensibility, the Cartesian isolation of the subject from the body is reversed and the subjective sensations are extended beyond the boundaries of the body into the huge expansion of the visual and auditory space, which now *incorporates* these modes of sensibility. When Descartes, in his answer to princess Elisabeth's question, admits that in the realm of sensibility the soul is connected to the body (and that he has strategically suppressed this fact in favour of pure thought), this admission forms the starting point of this reversal, which ultimately has an incredible consequence, that is, the assumption of an all-encompassing aesthetic subject—instead of one that is completely isolated from the body and extra-bodily phenomena. However, this epistemological reversal not only affects the Cartesian philosophy, but also the post-Cartesian philosophies, especially the doctrine of Kant's *Transcendental Aesthetic*, in which the apriorization of the sensibility initiated by Descartes came to fruition. With regard to the exteriority of space (as opposed to the interiority of time), as established by Kant himself, the doctrine of the transcendental aesthetic should primarily relate to the immediate space of seeing and hearing. However, in favour of his fundamental concept of the apriority of space (as "reine Vorstellung a priori"), Kant does not favour an aesthetic space, but a theoretical space—that is, the space of geometry, which is obviously an a priori spatial concept. The *natural* geometry of visual space, represented by its perspectivity and—even more clearly—by its geometrical-optical or dioptric conformity to the laws of reflection and refraction, is not addressed. This oversight does not appear to be a mere coincidence, but rather a strategy, because Kant builds the profoundest foundation of a transcendental subject and thus his doctrine of *transcendental philosophy* on the entirely a priori nature of space and time.

Strictly speaking, Kant's transcendental philosophical system is based more on a logical than on a pre-logical and purely aesthetic subject. Even when this system extends to the domain of aesthetic (the theory of sensibility), it is still limited to the apriority of a theoretical space that serves as a basis for geometry and the geometrical-mathematical sciences such as mechanics and optics. By contrast, the spatiality of the immediate sensory perceptions does not prove to be a mere a priori conception, but to be *given* and therefore capable of being a posteriori experienced. In *Transcendental Logic*, the epistemological distinction between the givenness of objects in empirical intuition and their a priori conceptual cognition—which is, however, conceived in the mode of a *synthetic unity of apperception*—is made very clear. Accordingly, the given factum of the object has a substantially different legitimacy within the framework of *transcendental logic* than the objects of sensibility, which is to be treated within the doctrine of *transcendental aesthetic*. Sensibility (Sinnlichkeit) does not belong to conceptuality (Begrifflichkeit) per se; therefore, it is inappropriate to include the factum of subjective-conceptual knowledge as an object of investigation in the doctrine of a transcendental aesthetic, which precedes the doctrine of transcendental logic and therefore only investigates sensibility. When the doctrine of transcendental aesthetic deals with geometrical or mechanical knowledge, it is fundamentally acquired in a purely visual mode—as visual knowledge. The factum of the object—of the body and extra-bodily objects—which is directly involved in sensibility, gives the sensory perceptions a real and not merely a priori imagined spatiality. However, this also applies to the temporality of sensory perceptions, which seems to be *given* by the object, both in the mode of the duration of its static state and in its dynamic movement. If our investigation is premised on the undeniable temporality or temporal state of phenomena, the a priori conception of time—even in the context of a transcendental-philosophical system—would be a subjective temporalization of *objectively given* temporality.[13] A detailed discussion of time as a purely objective and, as such, completely autonomous phenomenon from a transcendental subject could be the subject of a treatise ensuing this investigation.

References

Descartes, René (1965). *Die Prinzipien der Philosophie.* Artur Buchenau (Trans. & Ed.). Meiner Verlag. Hamburg.

Descartes, René (1972). *Meditationen.* Artur Buchenau (Trans. & Ed.). Meiner Verlag. Hamburg.

Descartes, René (2002). *The Principles of Philosophy.* John Veitch (Trans.). Blackmask Online (www.blackmask.com).

Descartes, René (2009). *Meditationen.* Christian Wohlers (Trans. & Ed.). Meiner Verlag. Hamburg.

Ebbersmeyer, Sabrina (2015, Ed.). *Der Briefwechsel zwischen Elisabeth von der Pfalz und René Descartes.* Wilhelm Fink Verlag. Paderborn.

Maier, Anneliese (1968). *Zwei Untersuchungen zur nachscholastischen Philosophie.* Rom.

Nietzsche, Friedrich (1994). *Die Philosophie im tragischen Zeitalter der Griechen.* Manfred Riedel (Ed.). Reclam Verlag. Stuttgart.

Ross, Colin A. (2010). *Hypothesis: The Electrophysiological Basis of Evil Eye Belief.* In: *Anthropology of Consciousness.* Vol. 21. Issue 1.

Ross, Colin A. (2011). *Traditional Beliefs and Electromagnetic Fields.* Revista de Antropologia Iberoamericana. Vol. 6,3. Madrid.

Thaliath, Babu (2001). *Der Prozeß der Geometrisch-Optischen Perspektivierung in der Visuellen Raumwahrnehmung.* FreiDok (Freiburger Dokumentenserver of the University of Freiburg). Freiburg i. Br.

Thaliath, Babu (2006). *Domänen des Subjekts.* FreiDok (Freiburger Dokumentenserver of the University of Freiburg). Freiburg i. Br.

[13] I have previously discussed the a priori subjective temporalisation of objective temporality in a paper entitled *Der Prozess der geometrisch-optischen Perspektivierung in der visuellen Raumwahrnehmung*—albeit in the context of the film aesthetics of Andrei Tarkovsky. See Thaliath (2001, p. 210 ff.).

MIX
Papier aus verantwortungsvollen Quellen
Paper from responsible sources
FSC® C105338

If you have any concerns about our products,
you can contact us on
ProductSafety@springernature.com

In case Publisher is established outside the EU,
the EU authorized representative is:
Springer Nature Customer Service Center GmbH
Europaplatz 3, 69115 Heidelberg, Germany

Printed by Libri Plureos GmbH
in Hamburg, Germany